Crisis Engineering

Time-Tested Tools for Turning Chaos into Clarity

Marina Nitze, Matthew Weaver
& Mikey Dickerson

balance

New York Boston

This publication contains the ideas and opinions of its authors. It is intended to provide helpful and informative material on the subject addressed, but is sold with the understanding that the authors and publisher are not engaged in rendering professional services in the book. The authors and publisher specifically disclaim responsibility for any liability, loss, or risk that is incurred as a consequence, directly or indirectly, of the use and application of any of the contents of this book.

Copyright © 2026 by Layer Aleph LLC

Cover design by John J. Custer.
Cover copyright © 2026 by Hachette Book Group, Inc.

Hachette Book Group supports the right to free expression and the value of copyright. The purpose of copyright is to encourage writers and artists to produce the creative works that enrich our culture.

The scanning, uploading, and distribution of this book without permission is a theft of the author's intellectual property. If you would like permission to use material from the book (other than for review purposes), please contact permissions@hbgusa.com. Thank you for your support of the author's rights.

Balance
Hachette Book Group
1290 Avenue of the Americas
New York, NY 10104
GCP-Balance.com
@GCPBalance

First Edition: April 2026

Balance is an imprint of Grand Central Publishing. The Balance name and logo are registered trademarks of Hachette Book Group, Inc.

The publisher is not responsible for websites (or their content) that are not owned by the publisher.

The Hachette Speakers Bureau provides a wide range of authors for speaking events. To find out more, go to hachettespeakersbureau.com or email HachetteSpeakers@hbgusa.com.

Balance books may be purchased in bulk for business, educational, or promotional use. For information, please contact your local bookseller or the Hachette Book Group Special Markets Department at special.markets@hbgusa.com.

Library of Congress Cataloging-in-Publication Data

Names: Nitze, Marina author | Weaver, Matthew (Founder of Layer Aleph) author | Dickerson, Mikey author
Title: Crisis engineering : time-tested tools for turning chaos into clarity / Marina Nitze, Matthew Weaver, and Mikey Dickerson.
Description: First edition. | New York : Balance, 2026. | Includes bibliographical references and index.
Identifiers: LCCN 2025049017 | ISBN 9780306836862 trade paperback | ISBN 9780306836879 ebook
Subjects: LCSH: Crisis management
Classification: LCC HD49 .N58 2026 | DDC 658.4/056—dc23/eng/20260121
LC record available at https://lccn.loc.gov/2025049017

ISBNs: 978-0-306-83686-2 (trade pbk.), 978-1-5387-8195-1 (special edition), 978-0-306-83687-9 (ebook)

Printed in the United States of America

LSC-C

Printing 2, 2026

Praise for
Crisis Engineering

"This is the book I wish every single boss I ever worked for had on their desk."

—Dan Davies, cyberneticist, author of *The Unaccountability Machine* and *Back of Mind* Substack

"*Crisis Engineering* is a field guide for anyone trying to make change with urgency, humility, and heart."

—Jen Pahlka, author of *Recoding America* and former U.S. deputy CTO

"This book gives you the tools to recognize a crisis, steer out of it, and maybe even make lasting change while you do. There's so much insight here: plenty of theory and useful new ways to think about crisis, but also practical tactics and detailed real-world case studies. (And the nerdy sidebars are fascinating.)"

—Tanya Reilly, principal engineer and author of *The Staff Engineer's Path*

"*Crisis Engineering* is the essential guide to facing the complex breakdowns that characterize the twenty-first century. Read it, and come away wiser, stronger, and readier to understand and deal with whatever comes at you."

—Tim O'Reilly, founder and CEO of O'Reilly Media and author of *WTF? What's the Future and Why It's Up to Us*

"The central fact of our era is the rot and stagnation of our institutions. Many words have been spent documenting and lamenting this state. *Crisis Engineering*'s contribution is a theory of how to get out of it, focused on the moments where brittle systems break. It's written by doers for doers."

—Patrick McKenzie, host of *Complex Systems*

"*Crisis Engineering* is required reading for every decision maker in a complex company, nonprofit, school, or government organization. This irresistible blend of theory, practice, and storytelling will help you avoid, reduce the duration and damage, and above all, take advantage of such unexpected, disorienting, and disruptive failures to make rapid system-wide improvements."
—Robert I. Sutton, Stanford professor emeritus and *New York Times* bestselling author of eight books, including *The No Asshole Rule*, *Good Boss, Bad Boss*, and (with Huggy Rao) *The Friction Project*

"The world is never going to get less challenging, systems won't get less complex, and the stakes won't get lower. What's more, the better you are at what you do—the more important and reliable your system is—the worse it will be when a crisis hits. This book gives you the understanding and tools you need so that, when your organization is suddenly in the most demanding moments of its existence, you can act with confidence, resilience, and grace."
—Deb Chachra, author of *How Infrastructure Works: Inside the Systems That Shape Our World*

"*Crisis Engineering* brings together insightful definitions, useful (and gripping) anecdotes, and decades of collective experience in the hot seat to help you not only manage and recover from a crisis, but to come out better."
—Jason Fraser, impact strategy advisor and coauthor of *Farther, Faster, and Far Less Drama*

"I have worked closely with and relied on *Crisis Engineering*'s authors to help solve crises few others would touch. The methods described in this engaging and helpful book will help you handle the tough problems. Read it now, before you need it."
—Denis McDonough, former secretary of the Department of Veterans Affairs and chief of staff to President Obama

To Wag Dodge and his escape fire

Contents

SECTION 0: OPENING MATTER
0.1: Introduction 3
0.2: What Is a Crisis? 11
0.3: What Is a Complex System? 19
0.4: What Is Crisis Engineering? 26

SECTION 1: TESTING THE THEORY
1.1: Sensemaking 101 31
1.2: Three Mile Island 42
1.3: HealthCare.gov 73

SECTION 2: YOUR CRISIS TOOLKIT
2.1: Crisis Engineering Toolkit: An Overview 97
2.2: Establish a Crisis Engineering Center 103
2.3: Map the System 119
2.4: Find Your People 144
2.5: Take Novel Actions 157
2.6: Manage the Story 184
2.7: Measure Progress 195
2.8: Communicate in a Crisis 207
2.9: When to Give Up 220

SECTION 3: EMERGING FROM CRISIS

3.1: Know When You're Done 225
3.2: Spin Down the Crisis 232
3.3: Instill Change in the Wake of Crisis 241
3.4: Plan for a Future Crisis 255
3.5: Engineer a Crisis Career 266

SECTION 4: CRISIS ENGINEERING IN ACTION

4.1: Ending California's Pandemic Unemployment Backlog 275

Conclusion *301*
Acknowledgments *303*
Appendix A *305*
Appendix B *309*
Notes *311*
Index *317*
About the Authors *327*

Section 0:

OPENING MATTER

0.1

Introduction

> If a man writes a book, let him set down only what he knows. I have guesses enough of my own.
>
> —GOETHE

You've found yourself in a crisis. No shame—everyone does, sooner or later. Systems are failing, customers are furious, regulators are circling, and nobody can agree on what's actually happening. The outage is spreading. The backlog is growing. Phones are ringing off the hook. You have minutes to act—hours if you're lucky—before the situation worsens. Taking the wrong action could cause a catastrophe, but if you don't take *any* action, you'll *definitely* cause one.

In moments like this, most organizations discover too late that their plans and instincts are no match for the speed and complexity of a real-world crisis. This is not just a guess. We have watched leaders of utmost experience and renown fumble the ball in some of the highest-stakes environments in the world: the White House Situation Room, the Department of Defense, and the boardrooms of Fortune 500 corporations. In the face of the unexpected and unknown, these experienced leaders did what's *usually* right: think long and hard, consult experts, make detailed plans, announce bold decisions, and spend billions of dollars. And yet despite these efforts, they still failed to resolve their crises. If their crisis did recede, its root causes remained in the shadows, ready to reemerge undetected at any time.

While stressful, and not most people's definition of a good time, a crisis is also a rare and powerful window of opportunity. In crisis, the normal resistance to change weakens, frozen structures begin to move, and previously impossible actions become possible. If you can act deliberately in this window—before it closes—you can transform your organization in ways that would otherwise take years, if at all. The goal of crisis engineering is to make this transformation *intentional*, not accidental.

This book is for leaders, operators, technologists, and public servants who find themselves responsible for navigating crises inside complex organizations, and who want to be ready to change the status quo when the moment arises. We are talking about system outages, major service disruptions, and any moment when reality changes faster than your existing structures can respond. Whether you work at a tech company, a government agency, a hospital, a utility, or anywhere critical systems run, you will face these moments. Knowing what to do and how to act with limited information, and while emotions are running high, is the difference between spiraling failure and transformational recovery.

Crisis engineering is the art of harnessing the process of sensemaking to not only successfully emerge from a crisis, but to leverage the crisis conditions into positive, enduring change in your organization. **Sensemaking**[i] is the automatic, continuous process that determines most decisions and actions taken by a person or a group. It's how humans reconcile and ascribe meaning to conflicting information. You can engage in purposeful sensemaking, but it's not default behavior, so you'll have to learn how. We'll show you.

Humans are an indispensable part of every complex system. They rarely question their habituated behavior, which makes it very

---

i If you're familiar with Daniel Kahneman's work, he referred to it as "System 1 thinking."

hard to change. A crisis is a unique opportunity to break out of this stasis.

Crisis engineering sounds technical and certainly applies to problems related to computers. But most of the time, even "computer" problems have causes (and solutions) that do not involve an electrical plug or an internet connection. Crisis engineering draws lessons from avalanche rescue, fire management, and hurricane response. It can be applied to a crisis of any kind and of any magnitude.

A crisis engineer finds points of leverage, balances the needs of a diverse portfolio in a rapidly changing environment, and operates at multiple time scales at once. They know how to find and validate ground truth, and can bracket and reframe information as needed. A crisis engineer understands that sensemaking is happening constantly, *whether they want it to or not*, and therefore confidently prefers nonintuitive behaviors like "guess and check" over "three-week planning session." This enables them to move faster and, in certain conditions, work at 100x leverage during the crisis (and perhaps residually 2x leverage the first Monday after the crisis has passed).

The Marines have a saying: "All bleeding stops eventually." Your crisis will end no matter what you do. Maybe the news cycle moves on to something more exciting. Maybe you scapegoat somebody and the press is satisfied by a head on a stake. Maybe you redefine "backlog" in such a way that you change nothing yet declare victory. Maybe it burns out on its own. Maybe you applied more duct tape and it's holding…for tonight. But it takes a crisis engineering approach to actually resolve the root causes of a crisis and emerge stronger as a result.

We wrote this book because we need more crisis engineers. The world's governments, institutions, businesses, organizations, and climate are experiencing unpredictable change at a rate and scale unprecedented in recorded history. Legacy systems and "tried-and-true" methods are failing to adapt to new circumstances. The resulting crises are expensive. A large system flailing in a crisis usually means a large number of people are suffering.

It doesn't have to be this way.

In these pages, we will:

- Show you how to identify crisis conditions, resolve the immediate crisis, and use the situation to produce lasting change in organizational behavior and culture
- Explain sensemaking in-depth so you understand what it is and why it is so integral to crisis engineering
- Review the stories of two notable crises—Three Mile Island and HealthCare.gov—to show you where sensemaking broke down, and how proactively restoring it would have made (or did make) a difference
- Provide a crisis toolkit for standing up a crisis engineering effort in your organization, including roles and responsibilities, communicating effectively, how to get and assess accurate information, and setting key health metrics
- Show you step-by-step how to sensemake by taking novel action, with many real-world examples

WHY US?

We've been at the forefront of many crises you've heard of, and many you haven't. And we aren't just practitioners; we also study and teach crisis engineering, both in executive workshops and in a college setting.

There are three of us: Marina, Mikey, and Weaver. Mikey and Weaver started as Site Reliability Engineers (SREs) at Google, where they were responsible for the operations of both web search and ads. Mikey was on the cover of *Time* magazine for his work running the HealthCare.gov rescue, for which he recruited Weaver to join him. Both moved into the federal government afterward to cofound the United States Digital Service (USDS). Marina met them there in 2014, when she was the chief technology officer (CTO) of the U.S.

Department of Veterans Affairs—an organization known for a crisis or two.

After we left the Obama administration, we (along with our colleague, Carla Geisser) started a crisis engineering firm called Layer Aleph, meant to help fix or manage complex systems. In those consulting jobs, we have taken a microscope to systems including Medicare, unemployment insurance, COVID vaccine distribution, wildfire recovery, and the back-office workings of a utility company and a multinational bank.

We have spent many hours intervening on very bad problems in very complex systems, and as principals charged with building and leading small teams of people to do the same. We've had front-row seats to nearly one hundred complex systems over the past two decades, with budgets ranging from the tens of millions to more than one trillion U.S. dollars. The problems in those systems have affected anywhere from tens of thousands to hundreds of millions of lives.

We have seen crises of every magnitude, and are convinced that the patterns and solutions are the same at any scale. If we hadn't seen it up close, we might still be laboring under the belief that when things *really* matter, there's a room somewhere full of grown-ups who know what they are doing. If you learn only one thing from this book, let it be this: There is no other room, and there are no grown-ups. No one is coming. It is up to us.

WHY YOU?

Odds are, you don't see yourself as a crisis engineer yet, because we just made those words up. But there is some organization and system you care about. You have probably noticed that there are times when it's unstable, and other times when it stubbornly refuses to stop doing something harmful.

We aim to help anyone who grapples with such a system. "Help" may take several forms. Depending on your role, it could be tactics you can apply directly, language you can use to steer an organization in the right direction, or an understanding of why certain problems are unsolvable and you should move on.

As fun as it would be to write multiple *Crisis Engineering* books, we are trying to help people in a variety of roles with this one. Here are some lenses that might make sense for different readers. Try them on and see what fits.

Executives

Well, you're an executive, so you're going to read the introduction and the conclusion. That's fine, we get it! There are deals to crush on this flight to Newark!

We kid. But yes, you can cover all our big ideas with these chapters:

- This introduction
- 0.2: What Is a Crisis?
- 0.4: What Is Crisis Engineering?
- 1.1: Sensemaking 101

Or skip to the toolkit entries that are most important for upper management to understand:

- 2.2: Establish a Crisis Engineering Center
- 2.4: Find Your People
- All of section 3: Emerging from Crisis

You can expect us to help you look and sound more prepared and confident the next time a fire breaks out, which is half the battle. We also detail exactly what you can do and say to get the right people engaged in crisis engineering, set them up for success, and get back to normalcy as fast as possible. After we've set you up for success, don't forget to buy many, many copies of this book for your leadership and technical staff.

Senior Technical Leaders

If you're a CTO, a principal engineer, or an engineering director, then you know that crises happen, and that they're likely to fall on your shoulders. We humbly suggest reading the whole book. Feel free to skip the case studies on Three Mile Island and HealthCare.gov if you find them boring. They are meant to illustrate crisis behavior that you have probably seen in real life already. The toolkit may also be familiar, but it should give you some new things to try, or suggest modifications to things you already do.

You can expect to learn new language that lets you communicate more precisely. But best of all, many of our readers in technical leadership positions report that this book has given them a new framework and tools for understanding behaviors they have already observed in their organizations. Understanding the underlying mechanisms clarifies thinking, answers some of the "why" questions, and has helped them design more effective solutions for their particular organization.

Individual Contributors

If you're a small cog in a big organization, you may not be called on to organize the next crisis engineering effort. Still, our ideas about why crises behave in predictable ways (section 0 and chapter 1.1) may be enlightening as you cope with whatever is happening around you.

The tactical chapters in sections 2 and 3 will be there when you need them. You may have opportunities to suggest improvements, and the day will come when you are in charge. Chapters 2.4 (Find Your People) and 3.5 (Engineer a Crisis Career) are relevant to you right now.

After reading, you can expect to be better prepared to contribute to a crisis engineering effort and to position yourself as a valuable member of a crisis engineering team. You may also now be able to nudge a disorganized crisis response in that direction. You can expect a better understanding of and ability to make sense of events as they unfold. Even if that sensemaking doesn't leave your own head, it will give you greater peace of mind, which is no small thing. We can also help you develop another underrated skill: how to recognize when further effort is futile and the best thing to do is to walk away.

0.2

What Is a Crisis?

> Long ago a science teacher told me, "The universe, she is a bitch." Several times since, I have thought about this sentence. It's probably right.
> —**NORMAN MACLEAN,** *YOUNG MEN AND FIRE*

In a large organization, rapid, directed change is possible only in particular windows of opportunity. We refer to these windows as crises.

When we first set out to define crisis engineering, we immediately encountered a problem with nomenclature. "Crisis" evokes bad feelings. Organizational crises are generally quite unpleasant, *but they don't have to be*. We're going to argue they have a bad reputation because they are generally badly managed.

Many people pointed us to the Chinese characters 危機, which are variously translated as "danger," "opportunity," "disaster," "crisis," or "change point." Taken together, they render as "Every crisis contains an opportunity" or "My enemy's danger is my opportunity" or "Danger is the time of change." The ambiguity encourages an expansive view of situations that have different combinations of danger and opportunity.

Then there is the ancient Greek κρίσις, which for centuries referred to the point in time at which either a fever would break or a patient would succumb to disease. It's too bad this usage is no longer in fashion, because it indicates that things may get worse but can also get better. That's pretty close to "a window of opportunity that permits rapid, directed change."

It seems a little too contrived to create a new word, so we will do the best we can with English. Thus, the word we are going to use is "crisis."[i]

THE FIVE CRISIS INDICATORS

It is possible to wield and direct the heat of a crisis toward desirable change, if you understand how and when. At the same time, some of the actions and models that work best in a crisis are useless, or worse, when not in crisis. The very first step in successful crisis engineering is the ability to tell whether you're in a crisis at all.

A ripe crisis will exhibit most, if not all, of these five indicators:[1]

1. **Fundamental surprise**
2. **Failure of sensemaking**—perceptions break down, existing maps and models don't work
3. **Degradation, disruption, or complete change of core processes or outcomes**
4. **High visibility**—either internal to an organization, external to it, or both
5. **Rigid deadline** or timeframe

Each makes it possible to take new actions and create new behavior.

---

[i] We know the word "crisis" causes particular problems for readers who speak the government dialect of English. The unwritten rules of government communication say that the word "crisis" can be applied only to something on the scale of war or a hurricane. Even then, a situation may be called a "crisis" only if it is over, if it was resolved in some positive way, and if we are crediting the heroic leadership of the current governor/president with the solution. If you are unable to set aside this definition of "crisis," please substitute any word you like. We'd give you the translation in government English, but we can't because that language has no words for unflattering situations.

Fundamental Surprise

Any organization's information-gathering, decision-making, and operational processes will work under most circumstances experienced *in the past*. Sometimes they even cover a certain space of unprecedented circumstances and inputs. A **fundamental surprise**[ii, 2] is an event or circumstance that violates the basic assumptions of an organization's consensus reality. **Consensus reality** is the current belief of how things work, what is happening, how we are related to the people and systems around us, and why we are here.

An example from our own experience rescuing the HealthCare.gov website: "You are the Department of Health and Human Services, the second-largest federal agency in the United States, largely in the business of calculating and sending checks to citizens, doctors, and hospitals. Legislation orders you to build an online health insurance marketplace—something you have never done before."

The Opportunity: Fundamental surprise forces people to accept new facts about reality.

Failure of Sensemaking

In normal times, all organizations operate under a certain amount of self-delusion. A sure sign that a crisis has begun is when those illusions become unsustainable. Either information-gathering has broken down entirely, or the disconnect between expectation and reality has become too great for anyone to ignore.[3] Channels for information flow can become useless noise. The rate of change can become so fast that information is too old to be useful by the time it arrives.

---

ii Other terms include "cosmology episode" (Karl E. Weick, "Collapse of Sensemaking in Organizations: The Mann Gulch Disaster," *Administrative Science Quarterly* 38 [1993]: 628–652) or "outside-context problem" (Iain M. Banks, *Excession* [Orbit Books, 1996]).

Critical sensory information like feelings, intuition, or context can disappear from operations or decision-making.[4]

A famous example of the breakdown of sensemaking was the Mann Gulch wildfire in Montana in 1949. Multiple mishaps and hazards conspired to undermine the fundamentals of the sensemaking process. The radio broke. High winds made it impossible to hear. Well-tested assumptions about fire behavior were violated. In the end, the team was scattered and trapped on an inescapable ridge, where all but three of the firefighters died.[5] The tragedy transformed Forest Service firefighting protocols.

Author Norman Maclean (who wrote *A River Runs Through It*) became fixated on the Mann Gulch fire toward the end of his life, as documented in *Young Men and Fire*—the source of most of the epigraphs in this book. We will return to this story often.

The Opportunity: Failure of sensemaking creates a chance to build a new, shared, and more accurate understanding of what is really going on.

Degradation, Disruption, or Complete Change of Core Processes or Outcomes

The least ambiguous indicator of a crisis is the degradation, disruption, or replacement of an organization's primary function. An organization's primary function stops, or is supplanted by some novel need. This is easiest to perceive in service organizations, infrastructure organizations, or platform providers. System failures ground all of an air carrier's flights. An intruder freezes a business's technological infrastructure. An automobile manufacturing behemoth must switch to building tanks. A bureau stops processing child welfare cases or unemployment claims, generating backlogs that grow without bound. A demand for a company's product or service increases by two orders of magnitude in a short time. The organization has an inescapable need to accomplish some urgent task, which may be novel, and requires behavior outside its historical norms.[6]

Core disruption has a direct effect on decision-making and how an operator in the situation might take action. A fundamental disruption of an organization's primary purpose drastically changes the impact of decisions. The consequences of decisions can now go far outside the usual range. The scope of a decision's outcome can change entirely. Over just a few minutes in the Mann Gulch crisis, the stakes of an individual smoke jumper's decisions escalated from "This fire could take a few extra hours" to "I may not survive." Most did not.

The Opportunity: Broken core functions mean that meaningful change is no longer optional; it's now required for continued existence.

High Visibility

Increased visibility, either in scope or intensity, often accompanies crisis circumstances. It can be perpetrated by regulators, auditors, middle management, employees, media, and/or whistleblowers. It increases the pressure on decision-makers, operators, and a majority of the members of an organization. Awareness of widespread attention often slows decision-making, reduces the quality of internally reported information, and increases fear. Ironically, outside attention reduces appetite for risk at a time when the only hope of success is trying something new.

Canonical signposts of high visibility include a small number of topics (perhaps only one) dominating internal communications, or the organization being featured in the news on a weekly or daily basis.

> *The Opportunity: Increased visibility can force action, as the price of inaction grows unacceptably high. Decisions can have much greater impact.*

Rigid Deadline or Timeframe

Most deadlines—and many so-called timing constraints—are more flexible than they appear. In practice, missing a deadline or product launch usually results in minor costs, a few extra meetings, or some internal finger-pointing—not a crisis. Even the IRS delays tax deadlines rather regularly. But some deadlines are truly fixed or nearly impossible to move. These kinds of constraints can be a significant contributing factor to a crisis.

Often, static deadlines result from physical constraints, as in aerospace applications, where launch windows are dictated by physics. They can also be imposed by higher-order control systems like courts, regulations, auditors, or financial systems. Sometimes previously negotiable timelines evolve into nonnegotiable ones, such as when progressing from a memorandum of understanding, to a draft contract, to a finalized and regulatory-approved merger, acquisition, or fundraising event.

Crises often impose nonnegotiable timing constraints that sharply reduce decision-making flexibility—for example, existential disruptions tied to immovable deadlines like a market opening or the close of an open-enrollment window.

Extant or imminent failure in the face of a hard, nonnegotiable deadline is a clear hallmark of a useful crisis.

> *The Opportunity: Rigid timing constraints mean that decision-making timelines are compressed. Matters that used to take months to decide will now be settled in hours or minutes.*

Not Everything Is a Crisis

Phony crises are much more common than real ones. There are many bosses who like to see heroic efforts, and try to generate them by turning everything into an emergency. There are also organizations where everyone, as a matter of culture, likes to pretend to be in crisis all the time. The case we get asked about most of all is when a bad situation is brewing from the bottom up: A system is failing and no one seems to notice.

In such a situation, look at actions, not words. Was there really a fundamental surprise? Have core processes stopped working? Is there an immovable deadline? If few or none of the crisis telltales are present, then crisis engineering isn't going to work.

We arrived at this list of indicators using existing research, lived experience, and experimentation. Each trait contributes in its own way to potentially creating durable positive change inside a sprawling and rigid organization.

Crises have the destabilizing power to shift incentives and invert power structures.[iii] They can radically alter priorities and disrupt existing understanding of complex systems. They tend to cause, or be caused by, a disconnect between the model of the world that exists in people's minds and the world as it is.[7] This is what makes sensemaking critical.

The destabilizing effects of a crisis happen no matter what. With or without active management, if the organization survives, the crisis will pass and the organization's incentives, priorities, and power structures will cool and solidify in a new shape. If the crisis does not pass, the organization will cease to exist in a recognizable form.

Without crisis engineering practices, a crisis is only likely to bring fear, decision-making paralysis, and further disruption to progress and change.

[iii] "It is easier to adjust actions than rules, easier to shift rules than change structures, and easier to alter structures than adopt new purposes. Only in the most drastic situations will it alter its fundamental purposes. Whatever adjustments it makes to the original change, it will eventually arrive at a new equilibrium point" (Anthony Downs, *Inside Bureaucracy* [Little, Brown, 1967]).

The good news is: If you can spot a majority of these circumstances around you, there is an *excellent* chance you are part of an organization primed for the willingness necessary for crisis engineering.

A future better than you were capable of imagining is now possible.

0.3
What Is a Complex System?

> The purpose of a system is what it does.
> —STAFFORD BEER

How can an organization made of people generate decisions and outcomes that no person wants? This commonsense rallying cry has led any number of reformers into battle against the system, with little success. Neither the most charismatic and empathetic managers, nor the cleverest and most efficient computer programmers, have much effect when they are up against the system.

When we say "the system," what comes to mind?

Every person reading this is a cog in dozens of complex systems. You may engage with some systems voluntarily, such as by enrolling your kid in youth soccer. More often, you have no choice but to engage with complex systems, such as when you show up to the hospital with a broken leg, or owe taxes to the IRS.

Computers have infiltrated all these systems.

A professor's job is to educate students. But most professors spend more time participating in committees, reviewing papers, filling out grant applications, reading and writing letters of recommendation, and other administrative tasks than they spend with students.[i] All of

---
i Citation: Mikey, who has taught part-time at Pomona College.

those tasks come to them via somewhat-automated computer processes, like email.

A doctor's job is to treat patients—as hands-on as jobs get. But they spend more time updating electronic medical records, fighting insurance companies, and doing other administrative tasks than they spend with patients.[1] All of these tasks are intermediated by somewhat-automated computer processes.

Even a plumber or a house painter must constantly maintain their representation in a dozen institutional databases, or they won't be operating for long. To operate a business, they must deal with professional licensing, accounting, payroll, multiple layers of permits, and several kinds of insurance.

Possibly the most system-forward jobs of all are bureaucrats and mid-level managers. Their jobs consist mostly of reading, updating, moving, or summarizing information in various computer systems, usually to compensate for the systems' shortcomings.

Whatever your own job is, it probably involves receiving information from some kind of computer system, doing something with it, and taking actions that are also in computer systems. You may never think of it that way, but that doesn't make it less true.

Across varying industries, all of these jobs have certain experiences in common. They have all had to stop and wait when the computers were down.[ii] They have all been frustrated wasting time relearning how to do something they could do yesterday, which has suddenly changed for no apparent reason.

Crucially, they have all seen a patient receive the wrong care, a plumbing job screwed up, a student unable to enroll in the right classes, or an insurance claim unfairly rejected, for reasons that seem to be nobody's fault.

These systems are misunderstood and underappreciated. Some new vocabulary and new analytical tools can help you get better and more predictable results out of them. The idea is to stop assuming you can manage the people as people and the computers as

---

ii Marina recently had to walk hospital staff through restarting their computer while lying on an operating table, in order for her procedure to begin.

computers, and instead consider both as part of the same complex system.

WHAT IS A COMPLEX SYSTEM?

When we say **complex system** in this book, we mean one that is a hybrid of human and machine components, that doesn't respond to simple interventions targeting only one or the other, and that generates unpredictable outcomes.

A complex system, in general, is a system that contains components interacting in ways that produce novel behavior. Novel (or emergent) behavior means that it could not have been predicted by looking at the subcomponents in isolation. Somehow, when they are assembled into a system, something more than the sum of the parts is created.

Some common examples:

- You can learn everything there is to know about a virus in a lab with a microscope, yet still have no ability to predict what will happen when an outbreak interacts with human hosts, social structures, and public policy.
- Predicting what will happen to a cubic meter of air under different conditions of pressure and temperature is the stuff of high school physics. Predicting what will happen in Earth's atmosphere tomorrow (let alone in one hundred years) is beyond human capability.
- If you own a corporation's common stock, you are entitled to a share of profits and you get to vote for directors. That's it. But when you put billions of shares into one place for millions of people to buy and sell, the resulting "stock market" is utterly unpredictable.

Complexity science is an enormous field of research, and we are going to do no more than use a few of its simplest findings. We really need only to borrow the insight that a complex system is one that can't be understood by decomposing it into parts, and we'll back away slowly and hope to avoid upsetting complexity theorists too much.[iii]

Complex Systems Are Fractals

It is possible to point an imaginary camera at almost any field of view and find a structure that can be labeled a complex system. It will have inputs and outputs that can be enumerated, and decisions that happen inside it. It can be evaluated as if it were a black box inside a larger system, or it can be taken apart into smaller systems. This sort of structure is called **self-similar**. Complex systems are **fractals**: a never-ending geometric pattern that repeats itself at different scales.

We discovered this reality long before we had any of the fancy words that we now use. Back in 2013, we (especially Weaver) referred to the HealthCare.gov system as a fractal of fucked-upped-ness.[iv] You could zoom in to any part of the diagram, and the amount of fucked-upped-ness in that subsystem was equal to the amount of fucked-upped-ness in the whole. We still consider this true, but today we would probably use other words in public. We'd also add another true statement: *All* complex systems are like this.

One exception to self-similarity is that the ratio of human to machine components is not constant up and down the stack. If you zoom in far enough on any part of the system, you are likely to find subsystems that are entirely automated. In our experience, the machine-only systems can be changed steadily and incrementally over time, so a crisis effort probably doesn't need to go any deeper than that.

[iii] Mikey found a series named *Foundational Papers in Complexity Science* and figured he would at least skim them to confirm we were using terms correctly, but he didn't anticipate that Volume 1 alone (1922–1962) would be six hundred dense pages.

[iv] There goes our PG rating. You may call it a fractal of dysfunction if your audience requires gentler language.

Directly outside the all-machine subsystems, which is to say in the lowest layers that include humans, you will often find processes that can be usefully automated. If you want to extend automation and reduce toil, this is the place to focus.

As one continues to zoom out, the density of machines decreases. Once humans have control of a process, it is not common for a machine system to be reintroduced as a means of regulating the humans.[v] Machines are unsuccessful at regulating humans because humans contain much more nuance.

It follows from the nature of fractals that a maximally effective crisis engineer will be a generalist: someone who can both fix a JavaScript bug and talk to the CEO. Conversely, working in the crisis engineering style will make you more of a generalist, quickly.

A useful way to think of "generalist" is that you are at least out of the basement on the Dunning-Kruger curve on any technical or social skill that might be relevant to your system.[vi] It's not necessary to be a world-class expert at every possible thing that can come up. If you know what you don't know, and can locate and ask for help, there will be nothing you can't solve.

What About Cybernetics?

We love calling things "cybernetic." The word is so out of date and out of favor that, like fashion trends, it's becoming cool again. It correctly and precisely describes the type of systems analysis that interests us.

"Cybernetics" was the name given to a field of research that sprang into existence, alongside the introduction of large-scale

---

[v] It is not impossible—some functions of the promotion and compensation systems at Google are algorithmic. The algorithmic decisions are then revised by human review and exception committees, reintroducing the complexity and variety.

[vi] If you are not familiar, Messrs. Dunning and Kruger gave their names to the phenomenon where people with the least competency in a particular area are also the ones who overrate themselves the most. The popular caricature is "If you are bad enough at something, you won't know you are bad at it." Try to avoid this.

computers, in the 1940s. Its better-known, surviving siblings are information science and computer science. Its successor and heir is probably operations research. Early cyberneticists were interested in a lot of things but settled on systems of automation and control, particularly those that have complex feedback loops.

Many of the observations survived and are in common usage, even if it's not possible to get a PhD in cybernetics today.[vii] The fact that we didn't have to define "feedback" in the previous paragraph is an example.

When we describe a system or a strategy as **cybernetic**, we mean that it relates to the complex interactions between humans and machines, and systems of control that are meant to regulate the same. For our purposes, we are borrowing two of the core insights from the field:

> **The Purpose of a System Is What It Does:** the mantra and Zen koan of Stafford Beer, the founder of management cybernetics. As he put it, "There is no point in claiming that the purpose of a system is to do what it constantly fails to do." This may sound like a pointless rhetorical barb, but when most people start trying to figure out a system, they do it with the unconscious assumption that the designers' intentions matter. *They do not.* The designers are gone—even if they are in the next cubicle, only their actions today matter. Enlightenment will come from focusing on the system, the actions, and the output as they actually exist today.
>
> **The Law of Requisite Variety**[2] says roughly that the amount of complexity, cleverness, and variety in any control system must be greater than or equal to the amount in the system it's regulating. It's hard to define precisely, but it's easy to see its effects. Many have noticed that the tire pressure monitoring system in their car fails more often than the tires.

---

vii Except in Australia. The Australian National University recently reestablished a "School of Cybernetics."

That's because it is much more complex, and it cannot be otherwise.

Learning the formal definitions and reasoning of the cyberneticists is fascinating, in part because they were such a bunch of oddballs. We can't fit it in this book. An excellent one-chapter overview appears in Dan Davies's *The Unaccountability Machine*. An excellent one-book overview is *The Heart of Enterprise* by Stafford Beer.

0.4

What Is Crisis Engineering?

Crisis engineering is the name that we made up for what we do. We have, through some fault of our own, been in the position of trying to fix or manage complex systems for a few decades. Through all our projects, we have tried to define success in terms of actual outcomes, rather than our level of effort or our ability to meet contract requirements. This has forced us to be rigid empiricists. We have tried many things that did not work. Do not be misled by the fact that the examples we use in this book all have passable outcomes. Our publisher wasn't interested in the pitch for the open-ended multivolume series *How Not to Do Crisis Engineering*. The things that worked repeatedly, that we think are important, became the much shorter book *Crisis Engineering*.

Our thesis, in short:

- A distributed system that comprises human and machine components is a special thing: a complex system.
- Just about every interesting problem worth solving today involves a complex system.
- The machine parts of a complex system can be modified slowly over time, but the human parts generally cannot. They only allow significant change in a short time window that we call a crisis.

- There are signature properties you can use to spot such a crisis.
- Sensemaking is the way to solve a crisis.
- Radical change happens when an organization negotiates its way out of a crisis using sensemaking.

You cannot avoid complex systems. You cannot avoid crises. Crises are opportunities to improve complex systems. The way improvements are made is sensemaking.

We're now in the middle-to-late phase of our careers, and we have standardized a lot of what we do. We teach workshops and college classes. In trying to find ways to communicate our experience, we backfilled the research and theory. We've discovered that some practices we found useful on the ground were already named and described in academic jargon years ago. We have cherry-picked our favorite papers and books from a half dozen fields, which eventually became the citations in this book. Some of our major influences include Daniel Kahneman and Amos Tversky (decision-making), Karl Weick (sensemaking), Donella Meadows (systems), Anthony Downs (bureau behavior), Stafford Beer (cybernetics), Nancy Leveson (safety engineering), and Joseph Heinrich (anthropology).

Our strategy does not solve all problems, but it does solve some, including some that are unapproachable by other methods. We hope it is useful to you, too.

Section 1:

TESTING THE THEORY

1.1

Sensemaking 101

> So this story is a test of its own belief—that in this cockeyed world there are shapes and designs, if only we have some curiosity, training, and compassion and take care not to lie or be sentimental.
>
> —**NORMAN MACLEAN**, *YOUNG MEN AND FIRE*

We will start with the core of successful crisis engineering. It is something invisible—something called *sensemaking*.

There is a force that surrounds you. It guides almost everything you do, and almost everyone around you. It has shaped the human-built environment that you live in. It caused you to be born and may well cause you to die. It's so ubiquitous that you have probably never thought about it or given it a name.

The force we're talking about is the relentless drive of all human brains to construct stories that make sense.

We have two disadvantages when talking about it. One is that it's so pervasive that humans can't see it, as in the joke about the fish who can't see the water. The other is that we can't think about how brains think without using a brain.

But even if we can't see a thing directly, we can deduce its properties. Consider oxygen: Even if you had no idea what it is, you'd notice pretty quickly that something wasn't right if you didn't have it. So to begin, we are going to talk about what sensemaking *isn't*.

Suppose you wrote down the process by which you make a decision. This question is intentionally vague. Generically, how do you make any decision at all?

You will probably come up with something like:

1. Determine your goal, or the problem to be solved.
2. Gather information.
3. Identify alternatives.
4. Evaluate the alternatives against some desirable criteria.
5. Choose the option that looks best against your criteria.

This is more or less your natural intuition for how you think you think.[i] And it's not remotely accurate.

Some of the first people to rigorously study decision-making considered how juries arrive at verdicts.[ii] Based on observations that started in the 1940s, sociologist Harold Garfinkel asserted:[1]

> *The outcome comes before the decision...Only in retrospect did they decide what they did that made their decisions correct ones. When the outcome was in hand they went back to find the "why."*

This is distressing to anyone invested in the idea that jury trials represent rationality and therefore justice. Others began looking for reasons to explain this behavior. They quickly stacked up a number of factors that seem to matter at least as much as the quality of evidence: trial structure, preexisting beliefs about the law, stereotypes, "demographic similarities," and the physical attractiveness of the players.[2] It's even been shown that judges make different decisions before and after lunch.[3]

---

i Part of this is culturally determined, of course. The list above describes vaguely Western-hemisphere, post-Enlightenment beliefs. People from different places and times might have included a step to pray for guidance, or consult an oracle, or get permission from a family elder. For our purposes, these variations don't matter.

ii Perhaps because there is a lot of money to be made in consulting on how to get the jury decisions you want.

A new theory was needed that could make this all make sense. In the 1980s, Nancy Pennington and Reid Hastie at the University of Colorado proposed the "story model." Their fundamental claim is that jurors impose a story on the evidence they get, and that intentions and cause-effect relationships are crucial to that story, *whether or not they are present in the evidence*. Later research showed that a jury's choice of story determines its verdict, and that choice is influenced by several properties of the candidate stories:

- Coverage: How much of the available evidence is explained by this story?
- Consistency: Does this story hang together, or does it contain internal contradictions?
- Plausibility: How consistent is this story with our general experience of other (real or imagined) stories in the world?
- Completeness: Does this seem to be the whole story? Or does it leave unsatisfying gaps?

Elsewhere, Daniel Kahneman at Princeton was getting disillusioned with the fundamental models in economics, which assume people make rational, self-interested decisions. It was clear that a lot of human behavior isn't rational. So, like the people studying courtrooms, Kahneman and his collaborators began with the hypothesis that there must be external factors that sometimes knock people off their rationality track. They named these cognitive biases, and you may have heard of some:

- **Confirmation bias** means that you tend to seek out information that reinforces your existing belief and discount information that contradicts it.
- **Hindsight bias** causes you to see past events as more predictable than they really were.
- **Anchoring bias** is when your guess at something quantitative and unknowable (like the value of a used car) is affected by someone else's asserted starting point.

- **Fundamental attribution error** is our tendency to assume that other people's behaviors are reliable indicators of who they really are. This is unlike the way we judge our own behavior, which we know is heavily influenced by our environment and current circumstances.

There are many cognitive biases, with new ones being added to the list all the time. It can be valuable to recognize these biases and have names for them when you see them playing out in a group. But at some point, when the list of exceptions and provisos gets long enough, it becomes reasonable to question whether the basic model of rationality was ever right. By 2011, emboldened by a Nobel Prize, Kahneman advocated his theory of two distinct decision-making systems, which he called System 1 and System 2.[iii] System 2 is the deliberative and careful machine for evaluating evidence in a rational way. System 1 decisions, however, operate on different rules. He describes System 1 as:

- Fast and automatic
- Effortlessly surfaces impressions and emotions
- Affected by associative memories, such as past experiences that resemble the current situation
- Running all the time; cannot be turned off

Kahneman goes on to argue that the vast majority of our day-to-day lives run on System 1 decisions—which is to say, on autopilot. We activate the slow and expensive System 2 only when forced. In fact, we will go to great lengths to *avoid* using it. This goes a long way toward explaining individual and group behaviors, as we will see.

Finally, there was a third school of academics developing similar conclusions, in the field of sociology. Karl Weick had been studying organizational behavior in crises. He defined **sensemaking** as *the automatic, continuous process that determines most decisions and*

---

iii It seems an editor decided they needed snappier names in the title of his book: *Thinking, Fast and Slow*.

actions taken by a person or a group. The idea is much the same thing as Kahneman's System 1, but the sensemaking theorists have constructed a more elaborate model of the process, which we find has a lot of power to predict behavior in crises. We mainly use Weick's theory and terminology from here on.

PROPERTIES OF SENSEMAKING

The effort to describe sensemaking has received book-length treatments of its own. In this book, we adapt the seven properties described in Weick's *Sensemaking in Organizations*:

1. **It's motivated by a sense of one's own identity.** Our need to understand what's happening around us originates in our need to contribute to the tribe.
2. **It's retrospective.** Past events inform sensemaking—plans and promises do not.
3. **It is an active process of cocreating your environment.** If you are a passive observer, sensemaking will happen slowly at best. It requires taking actions that change the environment.
4. **It's social.** Sensemaking is faster in a group, because we respond to unconscious cues such as the amount of surprise we see other members attach to a given fact.
5. **It's continuous, with no start or end.**
6. **It's built on facts that we select or deduce from the environment.** It always requires elevating some facts and discarding others.
7. **It's driven by plausibility more than accuracy.** We will prefer a familiar story that fits some of the facts over a weird one that fits all of the facts.

These are empirical observations—neither we nor the sociologists try to say much about why or how these things matter. But they *do* matter. If you change your environment in ways that cause "more sensemaking" to happen, chaos will recede, peace of mind will improve, and transformative changes will stick.

With decades of our own observations of organizational crises, we had developed our own opinions about things that work and things that don't. It was only when Mikey sat down to develop a college class that we looked around for any theoretical justification. We found that the "sensemaking" model fits our tools well: The things that work are those that play to one or more of the seven principles. Since then, we've looked to these principles for guidance when we're in a novel situation and unsure what to do.

THE ROLE OF SENSEMAKING IN A CRISIS

We're going to spend significant time in a book about crisis engineering talking about sensemaking, because the two ideas are inextricably related. The mechanics of sensemaking explains why organizations are only able to make radical changes in short windows of opportunity ("crises"). This pattern is what led us to sort out the occasional conditions that permit change versus the more common conditions that don't.

Mild incentives and logical persuasion are the sorts of things that economists—and most other people—believe should work to shape organizations. They don't. Theories built on the assumption that people are rational agents don't work well enough to predict any specific person's behavior, let alone plan around it. But people and organizations become fairly predictable if you start from the assumption that they are making most decisions, most of the time, on autopilot. The "System 1" brain, the one that runs an automatic, continuous sensemaking process—the one we rarely acknowledge even exists—is in charge most of the time.

The essence of what makes a crisis useful is that the convenient assumptions and shortcuts that maintain reality break down, creating a rare opening to craft a *new* reality that sticks.

Normal days contain minor surprises. You thought there

was milk left, but there isn't. Someone has parked in your parking space. You barely notice, because your brain contains invisible surprise-removing machinery that rationalizes surprises away. With only tiny injections of conscious thought, you construct a story that makes sense: Your husband probably used up the milk; a new employee didn't know how reserved parking works. There are inconveniences, but no crisis. Your normal day continues.

But your ability to process surprise isn't infinite. When there is too much surprise too fast, the machine jams. Suppose you woke up in a house you didn't recognize, with a person you didn't recognize, and stumbled your way to work, only to find it is now a Halloween store.[iv] In this scenario, most people would experience an uncomfortable feeling of not knowing what to do next, and of having lost their sense of place and identity. Even minor surprises would become unintelligible.

When it happens to one person alone, we call the resulting disorientation and frustration "midlife crisis" or "trauma." When a shock destabilizes a whole organization at once, that organization is primed to make previously unthinkable changes.

Sensemaking kicks into overdrive when we can no longer operate on autopilot. We simply can't think fast enough to get through an ordinary day on System 2 alone, let alone a day with novel challenges and time pressure. Having a broken-down System 1 brain feels stressful and scary. This motivates us to put it back in order.

If you have ever traveled to a sufficiently foreign country, you have experienced a mild version of this. Minor decisions like when to get up, what to eat, how to turn on a light, and how to get to dinner suddenly demand careful attention and reasoning. At their homes, Marina and Mikey know to arrive at the airport at least an hour before departure time, and what lines to stand in. When they were in Tuvalu, they found it hard to believe that the process was "wait at the hotel until you hear the plane land with your own ears, then

---

iv This thought experiment assumes that you don't often wake up in places you don't recognize, and don't work at a Halloween store. If you do, shine on: We're not here to judge.

walk over to the tarmac," so their System 1 brains made them go to the airport an hour early anyway.[v] Small tasks become difficult, and days that would have been dull at home are exhausting. You can't keep functioning in this way, so both brain systems work together to make sense of the new reality.

Put another way, **crisis**, as we are using the word, simply means that short period of time where the thinking part of your brain (System 2) is able to reprogram the sensemaking part (System 1). When applied to a system, it means the short period when the organization is able to design and install new behaviors.

There are only three possible outcomes after a system enters a crisis: The challenge is overcome with durable and deliberate change, the system dies, or the failure is rationalized into an accepted reality. The system will push itself toward one of these resolutions as fast as possible, because brains need to shift back into autopilot as soon as they can.

Rationalized failure is the most common result. Thus, most large organizations contain programs and departments that passively accept abject failure: infinitely long backlogs, hospitals that kill patients, devastating school closures that do little to affect a pandemic, and so on. These are fossils of past crises where the organization failed to adapt.

Avoid Dangerous Stories

In the rush to construct a new reality that accounts for the misfit facts, there are some common pitfalls. Any of the following stories is dangerous, because they are too simple to be true, and they will shut down further exploration, which won't fix anything:

1. **Blaming Individuals.** "It was Bob's fault." Every complex system failure has a Bob that made a mistake. Unfortunately, many

---

v They should have waited at the hotel. It was hot.

postmortem processes stop here. This rarely solves the problem at hand. Instead, it makes things worse by creating an atmosphere of fear that kills collaboration. Even if Bob isn't punished, ending an incident analysis with "human error" is a dead end that teaches nothing.[vi]

» If you're tempted to fire someone today, move past it by asking the question: Somewhere, Bobs make mistakes every day. Why was today's outcome different?
» You can always fire Bob later if we are wrong.

> The National Highway Transformation Safety Administration (NHTSA, pronounced like "knits a sweater"—conveniently, also how to pronounce Marina's last name!) was born of crisis, as Congress went after automobile manufacturers for record deaths and injuries on highways and demanded "something" be done. To eschew responsibility, the manufacturers pointed fingers at "the nut behind the wheel."[4] This blame game saved no lives. What did save lives was moving NHTSA's focus away from blaming drivers and instead onto making cars "crashworthy," with innovations like air bags.[5]

2. **Overtraining.** A very common step after a system failure is to expand the training manual from 2,000 pages to 2,007 pages. Will employees look at it? Will they retain the information? The answer is almost certainly no. Training must be practical, targeted, and aligned with how people actually work. The same goes for any other new information source, like another dashboard or another alert.
3. **The Legacy System Excuse.** "The mainframe did it." Blaming outdated technology is hardly better than blaming Bob. Most

---

[vi] This perspective inspired Sidney Dekker's work investigating airplane accidents. If the goal is to blame the pilot, you will not get the critical information necessary to make *systemic changes to make air travel safer overall.*

likely, the outdated technology was working fine until very recently. What changed?

» Ripping and replacing legacy systems is unrealistic, expensive, and disruptive. A more useful challenge is understanding and addressing the interactions between people, processes, and technology. All of Western civilization runs on mainframes;[vii] yours is not the problem.

4. **Things Are Just Too Complicated.** Yes, it's probably true that the system has accumulated a lot of complexity. It is useless to complain about it unless there is a realistic path to radical simplification. Usually there is not. Failing that, the operators are just going to have to saddle up and learn to cope with the system as it is.

"In their eagerness they suppose that announcing a rule designed to forbid whatever behavior led to the criticism actually will work. Their immediate subordinates, remote from field pressures (and perhaps eager to ingratiate themselves with the executives) will assure their bosses that the new rule will solve the problem. But unless the rule actually redefines the core tasks of the operators in a meaningful and feasible way, or significantly alters the incentives those operators value, the rule will be seen as just one more constraint on getting the job done (or, more graphically, as 'just another piece of chickenshit')."

—JAMES Q. WILSON, BUREAUCRACY

[vii] We have checked with our own eyes.

Convenient stories offer the illusion of resolution without addressing the systemic complexities that caused the problem in the first place. They do nothing to make an organization better off. Watch out for them.

Examining the Theory

In the next two chapters, we'll describe cases where the sensemaking process broke or slowed down enough for us to examine it in action. Afterward, we'll return to these sensemaking principles and argue how they were represented in the case studies.

Let us warn you now: The case studies are dense and complicated. We wish it weren't so, but our point is to show that real-life crises that emerge from complex systems have many causes and do not proceed in a neat, linear fashion. This is why the value is in mastering the sensemaking process, not any particular magic-bullet solution. There is no substitute for building up your ability to handle complexity.

Readers who don't want this exercise and don't require convincing can skip to section 2, which is a concrete toolkit for applying sensemaking strategies.

1.2

Three Mile Island

> In this story of the outside world and the inside world with a fire between, the outside world of little screwups recedes now for a few hours to be taken over by the inside world of blowups, this time by a colossal blowup but shaped by little screwups that fitted together tighter and tighter until all became one and the same thing—the fateful blowup. Such is much of tragedy in modern times and probably always has been except that past tragedy refrained from speaking of its association with screwups and blowups.
>
> —**NORMAN MACLEAN**, *YOUNG MEN AND FIRE*

There are few domains where mistakes are serious enough to lead to detailed, thorough investigations. Even rarer are accident reports that are made public. This is why anyone interested in how to manage an organization in crisis ends up reading about disasters in the nuclear power industry.

Now it is your turn. We will look at one of the most dissected crises in history: the 1979 reactor meltdown at Three Mile Island that effectively ended the American nuclear power industry.[i] And we will do so through the lens of sensemaking.

---

i Our account will omit many details and will be somewhat stylized, which was necessary to fit it into this book.

NUCLEAR REACTOR 101

The story of Three Mile Island benefits from a little background in how nuclear reactors work. Don't worry—you'll be able to follow just fine.

In a pressurized water reactor, the reactor core lives in a gigantic steel pressure vessel about forty feet tall, fifteen feet across, and shaped like a cigar. Inside is a delicate rack assembly that holds an array of thin metal tubes full of radioactive uranium fuel. The tubes are arranged at a density that is just right for the fuel to create a self-sustaining fission chain reaction that releases heat.[ii]

Everyone is familiar with a different self-sustaining chain reaction that releases heat: *fire*. The crude analogy here would be that the fuel rods act like charcoal bricks in a grill. They burn better and hotter in a group than they do on their own, and the arrangement matters. Unchecked, they will burn until all the fuel is consumed. Uranium is different from charcoal in one way: The "fire" (radioactive decay) is *always* present at a low level. Nothing needs to ignite it. When enough fuel gets close enough together, the heat-producing reaction accelerates all on its own.

So we have a machine that will heat itself up to practically unlimited temperatures, capable of destroying itself if it passes the melting point of its various materials. That's fun, but not very useful. We need to carry that heat out of the reactor core and use it to generate steam, because we know how to use steam to generate electricity. We convert the reactor into a water heater by circulating water through the core. Water carries the heat to somewhere more useful, which also keeps the whole apparatus from melting itself.[iii] But the water comes out radioactive, so we want to keep it contained in a closed loop. We will call this the **primary cooling loop**.

---

ii The array also contains control rods that can be inserted or withdrawn to control the speed of the reaction. We're ignoring them because, ironically, they do not factor into the Three Mile Island accident.

iii The water is performing a third critical function, which is that it is a neutron moderator, making the sustained chain reaction possible. This also doesn't matter for our purposes.

The primary cooling loop is easier to control if we keep it full of liquid water (no steam). You could imagine a reactor that ran at low enough temperatures for the water to stay liquid at normal pressure. But this is America—we want to make a lot of power, so we run the water much hotter. This is fine as long as the water is kept under high pressure. We add a pressurizer, which is a reservoir attached to the primary cooling loop, to modulate pressure by adding or removing heat.

The pressurizer has one other function that factors into our story: It is the one place in the primary cooling loop where a steam bubble is allowed to exist. This is because the air bubble is compressible, like a shock absorber, which protects the whole system against shock waves, called **water hammer**.[iv]

Now we're getting closer to a useful machine: It doesn't melt, and it produces hot, high-pressure water. All that's left to do is turn the heat into electricity. This step happens the same way as any other power plant. We call this part, which pushes steam through a generator, the **secondary loop**.

Schematically, this is all we need for a working power plant.

The overwhelming complexity of a real power plant is in the pesky operational details. You need a way to access the inside of the core to refuel. You need to move water in and out of the various reservoirs. Water picks up contaminants that you need to filter out. Everything is always very hot and under pressure, so materials handling gets complicated.

Another factor that increases the complexity of a real plant is that the operators want high reliability, and the strategy to achieve this is *massive* redundancy. Each "pump" that we described is really *many* pumps, check valves, and sensors. For each such array, there is a backup array. Backup pumps require backup power sources. And so on and so forth, until the control *panel* has become a control *room*, and looks like this:

---

iv This is the same reason that closed-loop water-heating systems in houses have an expansion tank. It will be the menacing-looking red pressure tank near and above the furnace.

NASA Plum Brook Station, Nuclear Reactor Control Room *(Yasmine Abulhab, Bloom Works)*

THE ACCIDENT AT THREE MILE ISLAND

The night of March 27, 1979, was clear and a little colder than normal in Harrisburg, Pennsylvania. The operators of TMI-2, the newer of the two reactors at Three Mile Island, were working on some filters. They needed to be back-flushed, a routine maintenance procedure to clear clogs. It wasn't going well, and the operators were stuck.

Somewhere around the late-night shift change, there was either an operator mistake or a malfunction in one of the filter valves. Just after 4 a.m., it caused a water hammer, which damaged the inlet valves across all the filter tanks and stopped the flow of water through the whole secondary loop.

Things happened fast from here. With no water to pump, the secondary loop feedwater pumps, which push water into the steam generator, all shut down. Backup pumps automatically turned on, but their control

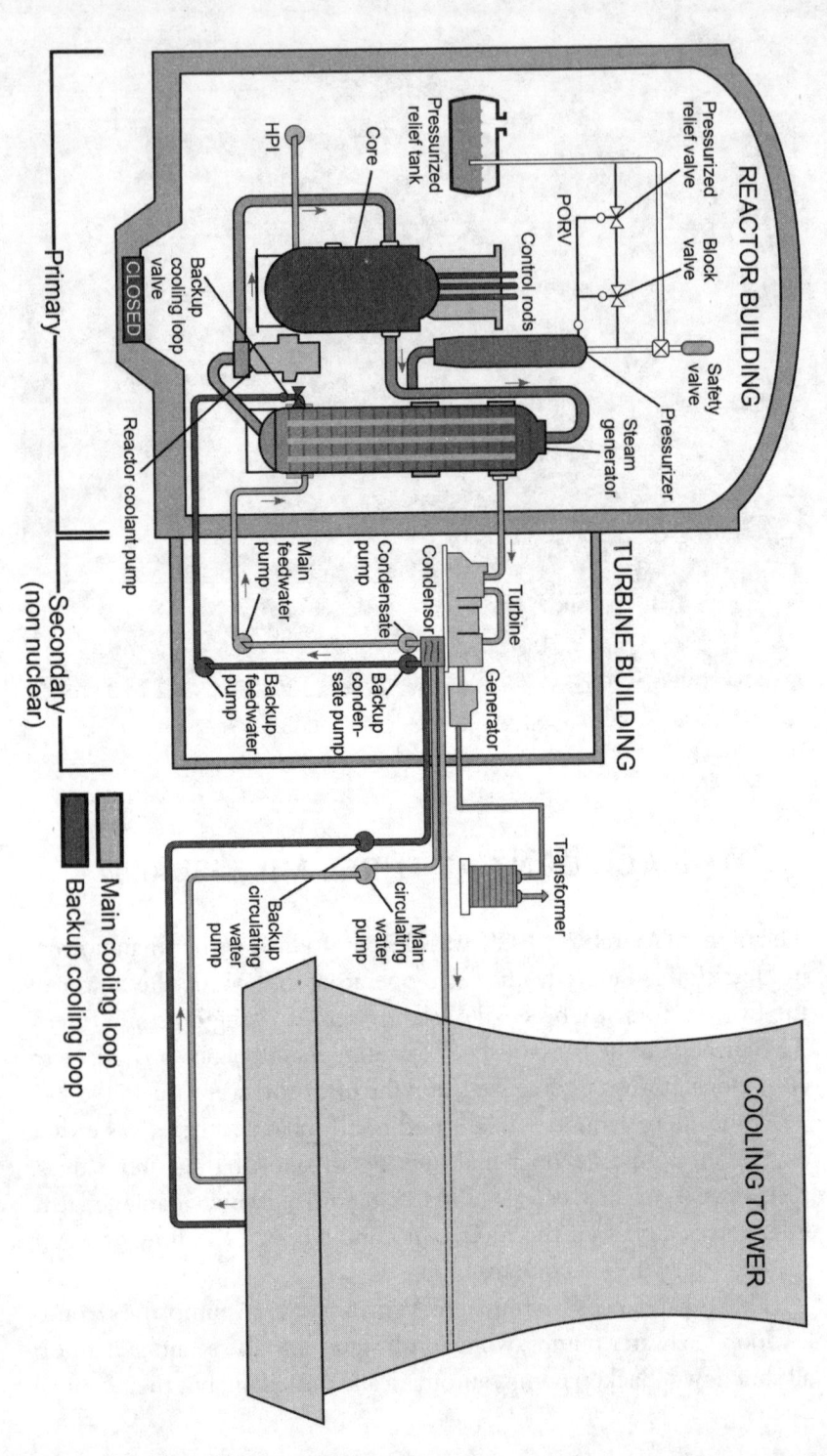

valves were closed,[v] so they could not move any water, either. With no water going in, there was no steam coming out, so the turbine automatically shut down. Loud alarms sounded in the control room, indicating multiple failures that all seemed to happen at the same time.

From the perspective of the local electrical grid, there was already an emergency. Grid operators have to maintain a close match between the amount of electricity being generated and the amount being consumed, to avoid brownouts and damage. TMI-2 had been supplying 879 megawatts of electricity (enough for about two hundred thousand homes) that suddenly disappeared.

The secondary loop carries the heat away from the primary loop, so when it stalled, the primary loop started to overheat. Very soon, the primary loop was 100 psi over normal, and the pilot-operated relief valve[vi] (PORV) at the top of the pressurizer vessel automatically opened.

Hot water and steam began to drain into a storage tank underneath the building. More alarms went off.

Inside the core, the fission reaction was still raging away, generating gigawatts of heat that had nowhere to go. The temperature and pressure in the primary loop continued to increase, even with the relief valve open. The last automatic failsafe engaged: The reactor control system executed an emergency total shutdown of the reactor core, known as a **scram**.[vii]

The time from the initial feedwater pump alert to the scram was eight seconds.

The scram worked, and the fission reaction in the core came to a stop. This left only the small matter of decay heat. The nuclear reaction that splits uranium atoms leaves behind unstable waste

---

[v] Later investigators guessed that they had been left closed after a maintenance procedure weeks before.

[vi] "Pilot-operated" refers to the way the valve works mechanically; it does not mean it is controlled by the plant operators.

[vii] The Nuclear Regulatory Commission explains the origin of this term: At the Chicago Pile (the first atomic reactor developed for the Manhattan Project), Volney "Bill" Wilson added a big red knob to the control panel. Wilson explained that you'd push it if there was a problem. When asked what to do after pushing it, he reportedly replied, "You scram out of here." The word appears to have stuck. (See Thomas Wellock, "Refresh—Putting the Axe to the 'Scram' Myth," U.S. Nuclear Regulatory Commission, https://www.nrc.gov/reading-rm/basic-ref/students/history-101/putting-axe-to-scram-myth.html.)

products. They decay at their own pace, releasing more heat, until they eventually reach a stable isotope. This means that after a reactor is shut down, it continues to generate heat at the level of a few percent of its most recent output.

When TMI-2 scrammed, the decay heat was an amount that the primary cooling loop could handle. After the scram, the pressure began to drop, and the PORV was no longer needed. On the control panel, it looked like the valve was closed.

In reality, the valve had jammed and was still open.

So far, the operators had thirteen seconds to react. They probably had barely enough time to shut off the alarm bells and sirens. They could hardly be expected to have read all the alerts and warnings, let alone make any sense of what happened.

Their first recorded action was to confirm that the backup feedwater pumps were running. They were. But the operators did not notice two other lights indicating the valves *ahead* of those pumps were closed. The operators believed that water was circulating through the secondary loop, when it was not.

The last cooling water boiled away after ninety seconds, and the temperature in the primary loop began increasing rapidly, like a dry pot on the stove. There were erratic pressure readings. The high core temperature activated the emergency core cooling system (ECCS), meant to keep the core underwater no matter what else has failed. Then the ECCS triggered one of its subsystems, the high-pressure injection (HPI) pumps, which injects cold water directly into the core at high pressure.

Two minutes passed, which must have included deliberation that was not recorded.

Then, the operators turned the HPI pumps off.

In reality, what's happening is: The secondary loop is stalled. All its water is gone. The primary loop still has water leaking out through the PORV, which is jammed open. The reactor core is generating decay heat.

But remember, the operators don't know the relief valve is open. Therefore, they believe both cooling loops are working properly and everything is back to normal.

The next events therefore made no sense to the operators. Pressure decreased and temperature stabilized, but the water level in the pressurizer was much too high. In a sealed system, all three variables would move together.[viii] The reactor seemed to be violating the laws of physics.

The pressure was low because of the leak, and the reactor core had started to boil. The instruments and sensors in the primary loop were designed under the assumption that it would *never* contain air pockets. They were therefore now unreliable. The now-uncontained water maintained its own equilibrium of temperature and pressure as it boiled away.

Five minutes had passed since the start of the accident, with only a few minutes left to change the reactor's fate. What the operators saw was stable temperature, stable pressure, and an unexplainable high water level in the pressurizer. Something weird had definitely happened, but the situation seemed manageable.

The operators, who had already shut down the HPI pumps, decided that the unnaturally high water was too dangerous. They opened valves to drain some of it, unwittingly dumping the last of the cooling water that would prevent a meltdown.

Three minutes later, at minute eight, the core was dry, it was still generating about sixty megawatts of decay heat in its confined space, and it melted itself.

In the subsequent hours, the operators and an increasing number of managers remained confused and disoriented. They did their best to manage the symptoms of the meltdown, such as the mechanical failure of the dry feedwater pumps and the mysteriously huge volume of drain water that overflowed the containment building.

From the perspective of an engineer who wants to learn how to tactically prevent this kind of accident, nothing that happened after minute eight mattered. But from a systems perspective, the consequences of the accident were mainly determined by management decisions after the fact. We will examine both.

[viii] The water level sensor is in the pressurizer. The pressurizer is above the reactor core. Thus, gravity being what it is, the operators are trained to assume the reactor core is full of water if there is any water in the pressurizer at all.

TIME	REALITY	OPERATOR PERCEPTION
04:00:36	Water hammer damage causes feedwater pumps to shut down. Backup pumps are turned on but cannot move any water because their lines are blocked by closed valves.	Feedwater pumps trip. Backup pumps start automatically.
+2 sec	The steam turbine and electric generator shut down automatically.	Lots of alarms and warnings go off at once. What they understand is there has been a turbine trip.
	Primary cooling loop pressure increases to 100 psi over normal.	
	Pilot-operated relief valve (PORV) activates automatically.	
	Hot water and steam begin to drain into a storage tank.	
	Pressure continues to increase.	
+8 sec	Reactor core automatically executes emergency stop ("scram").	There was a scram.
+9 sec	Heat production is stopped; only decay heat remains at 160 MW.	

TIME	REALITY	OPERATOR PERCEPTION
+13 sec	Pressure decreases to 50 psi over normal. The PORV is automatically deactivated, but it sticks open because of a mechanical failure.	According to the indicator in the control room, the PORV is closed.
+14 sec	The backup feedwater pumps are powered but not moving any water because the rarely used block valves are closed. The block-valve warning lights are partly obscured by a maintenance tag hanging off the control panel.	Operators confirm that the backup pumps are running.
+48 sec	In the primary loop, pressure is going down, and water level is going up.	
+105 sec	The last water in the secondary loop boils off, leaving it dry.	
	Primary loop temperature and pressure are increasing again.	
+2 min	Pressure suddenly drops dramatically. The reason is not known. Most likely, the rate of leakage through the broken PORV increased.	
	The emergency core cooling system activates. HPI pumps water into the core at high pressure.	

TIME	REALITY	OPERATOR PERCEPTION
	Water level is increasing, pressure is decreasing, and temperature stabilizes.	Pressure and temperature are OK, but the water level is too high.
	Up to this point, the automatic failsafe systems have all worked successfully. If no one takes any action, the core will reach a safe cold stop in about an hour.	
+4 min		Operators turn off the HPI pumps, leaving only one running at 10% power.
+5 min	Water starts to boil in the reactor core, forcing more water into the pressurizer.	The water level is even higher. Operators begin to drain water out of the primary loop.
	It's unlikely that any action taken after this point can prevent the meltdown.	
+8 min		Operators notice the secondary loop is not running, find the closed block valves, and open them.
	Water begins to circulate through the secondary cooling loop.	
+11 min	With the PORV still stuck open, the drain containment tank has overflowed and water is flooding the building.	New alarms indicate water in the containment building, where water should never be.

TIME	REALITY	OPERATOR PERCEPTION
+15 min	The drain storage tank bursts. Pumps start to transfer the overflow water out of the containment building into an auxiliary building.	Instruments for core reactivity, temperature, and pressure in the containment building are all abnormal and increasing.
+20 min		Operators turn on the cooling and fans in the containment building, not knowing why it is so hot in there.
+39 min		An engineer in an auxiliary building discovers it flooding with overflow water. He calls the control room. Operators shut down the sump pumps, not knowing where the water came from.
+45 min		The day shift superintendent arrives. He is told there was a turbine trip and a scram.
+60 min	All the water in the system has become steam, causing the feedwater pumps to violently shake.	The whole building starts to shake.
+74 min		Operators shut down two of the four feedwater pumps.[ix]

---

ix There is a discrepancy here in the report from the Kemeny Commission. If "feedwater pumps" refers to the backup pumps, there were three of them. If it refers to the main pumps, they must have been turned on again at some point that is not mentioned.

TIME	REALITY	OPERATOR PERCEPTION
+101 min		Operators shut down the remaining two feedwater pumps.
	It's now about 5:40 a.m. There are radiation alarms in the containment building. Several more managers have arrived in the control room.	
	The zirconium in the fuel rods is breaking down, creating explosive hydrogen that is collecting in the containment building.	No one yet understands that the primary cooling water has boiled off and leaked out through the stuck valve, and that the core is exposed.
+120 min		Conference call with four managers on it, including the day shift superintendent. Someone asks him to go check whether the post-PORV block valve is closed. The answer comes back "yes."
+122 min	The block valve is closed.	
	From here, the control room and the increasing number of managers remain confused and disoriented for several more hours. They will eventually declare a site emergency at around +180 min (7 a.m.). The reactor core has already melted, and nothing they could have done would have changed the outcome by much.	

WHY DID THE ACCIDENT HAPPEN?

The accident at Three Mile Island was a major influence on the "normal accidents" work of Charles Perrow, the "systems safety" work of Nancy Leveson, and just about every other study of complex systems after 1979. It is a great example of how looking for a "root cause" is not enough. Many factors worked in concert to produce a meltdown.

Mechanical Failures

Most obviously, there were mechanical parts of the reactor plant that broke or didn't work on the morning of March 29:

- The water filter clogged while operators tried to flush it.
- The equipment the operators had to clear the blockage (a pressurized air system) was insufficient and didn't work.
- The PORV failed to close when it was deactivated thirteen seconds into the accident.

Operator Errors

Nearly as obvious is the fact that some of the operator actions made the situation worse:

- They shut down the HPI system at minute four. Absent another malfunction, the HPI system would have kept the core covered for at least the critical first hour.
- They drained the remaining water in the primary loop at minute five, which was the opposite of what was needed. This brought the reactor to the point of no return.

- They did not enable the backup feedwater pumps in the secondary loop until minute eight. If the secondary loop had stayed up, the temperature and pressure in the primary loop would probably never have gotten out of control.

Most investigations stop here. The purpose of a congressional inquiry or a lawsuit is to assign blame in a way that is convincing to an unsophisticated audience—not to truly get to the bottom of things. The press and Congress mostly concluded that the accident was the fault of a bumbling and lazy power company. To the extent the power company was able to defend itself, those defenses only served to make nuclear technology look irredeemably dangerous, which was worse. Much of this happened because of poor communications, as we will see.

To a systems thinker, this is a useless sideshow. Mechanical parts are always going to break down, and human operators are always going to make mistakes. These factors were present on every day of the plant's existence up until now, and on all of those days, it did not melt down. Placing blame on individual people or machines results in a missed opportunity to reshape the big picture.

Plus, the factors that actually created this accident are more interesting.

Organizational Failures

Looking only at a dry account of the sequence of events, the operators' behavior makes no sense without some missing context: The people at the controls on March 28 were Navy veterans. Moreover, the engineering and training staff at Babcock & Wilcox (the reactor manufacturer) relied on a lot of Navy experience, as did the entire civilian nuclear power industry.

Navy training fundamentally changed the way the operators interpreted and responded to the event. U.S. Navy reactors—the ones that drive submarines and aircraft carriers—have important differences from power reactors. Though they use the same basic design, the naval reactors are comparatively tiny.[x] The residual decay heat of a power reactor *after shutdown* is greater than the output of a naval reactor *at full power*. Submarines also need to operate quietly. The air bubble in the pressurizer performs an important noise-suppression function.

In other words: On a submarine, losing that air bubble will get you killed, and an overheated reactor probably won't. Decades later, power reactor operator training continued to preach this dogma: "Never let high-pressure injection pump you solid."[1] The exact opposite is true for power reactors: Losing the air bubble is bad, but not anywhere near as bad as losing coolant.

In addition to the Navy training that pulled their attention elsewhere, the operators faced poor design choices that made it difficult to notice the loss of coolant.

Under normal operation, the sealed primary cooling loop contains mostly hot water, with a small steam bubble in the pressurizer, as noted. The plan for a "loss of coolant" accident assumed a water leak. If water is leaking out of the loop, then both pressure and water level will drop together. This is how crews are trained to recognize a leak. If the leak is at the very top of the system, above the water level in the pressurizer, then the water level will increase as pressure drops. At this very top is where the PORV sat.

Worst of all, there was no direct measurement of the water level in the reactor core at all. Only the pressurizer was instrumented, on the assumption that if the pressurizer contained any water, then the core must be full.

When the PORV jammed and the core started to boil, this design assumption no longer held. The system was now in a condition for which the operators had no training, presenting data they had never seen before. The only hope of avoiding the accident would have

---

[x] The S2W reactor installed in the USS *Nautilus* in 1954 was designed to generate 13,400 mechanical horsepower, or 10 MW. Reliable information about later designs is hard to come by, since it tends to be classified.

been some *very* quick improvisation to resolve the many instrument readings that surrounded them into a novel, correct, and sensible picture of the state of the reactor. They needed new insight on the spot, and they had eight minutes to develop it.

The design of the operators' environment didn't help. The control room contained thousands of controls, instruments, and warning lights, arranged according to no particular design principle. When the trouble started, dozens of alarms fired in a matter of seconds. When an operator receives multiple alerts at once, they get no clues as to which ones are causes and which are effects. There's a good chance of getting cause and effect wrong when there are two or three alerts; given a hundred, it could take a team of experts weeks to untangle.

Sensemaking was further impeded by some instruments that were not what they seemed. The PORV, for instance, had a green light and a red light. The manual might have defined them accurately, which is that green means "the *electrical signal* to close the valve is being transmitted" and red means "the *electrical signal* to open the valve is being transmitted." But it's hard to imagine that an operator, who has probably never even seen the red light outside of maintenance or a training simulation, is going to remember that nuance. They are more likely to interpret green as "good" (closed) and red as "bad" (open). That's likely what happened in this case: They assumed the green light meant the PORV was closed, when it wasn't.

At the same time, other indicators and instruments correctly revealed crucial anomalies but escaped notice. The valves ahead of the backup feedwater pumps were not supposed to be closed in normal operation, but they were. The status of the valve behind the PORV, which could have stopped the water leak, was never established with much certainty. According to testimony, a manager asked over the phone for someone to go check on it, and received the ambiguous answer, "It is closed." Of course, this was already two hours after the accident, so it no longer mattered.

Cultural Failures

Most organizational behaviors are produced by culture, which is to say, habits. The habit of focusing on the pressurizer led the TMI operators to make disastrously wrong decisions. This is a narrow point, even though it had large consequences in the accident. There are trends visible in the TMI organization that were likely to have wider effects. We'll deconstruct just two of them here.

FETISH FOR COMPLEXITY

A hugely critical and easily overlooked factor in this story is that the TMI human-machine interface was *not* designed for operators. The practice seems to have been to take each instrument and control and put it anywhere there was a space. Instruments were grouped by how they fit: all gauges on one panel, all sliders on another. Maybe it made the control room easier to build, or somebody thought it was more visually appealing. There weren't any chairs at most panels, because the relevant instruments for any particular task are scattered around the room; operators were accustomed to a lot of walking and coordinating across multiple people. Good for their daily step counts, bad for making sense of a dynamic situation.

Other industrial systems built in the same era are similar. They are systems with a lot of unavoidable complexity. Still, all of the control designers seemed to agree that more is more.

For a similar example outside of power plants, below is the original cockpit of a Boeing 747.

Complexity made these machines *look* more impressive. A very complicated-looking machine will attract more attention, command higher budgets, and inspire the appropriate degrees of public awe. There is a reason why most of these control rooms had a press gallery.

These are the incentives that dominate the "novelty" phase of a new technology. A bigger, more expensive, more complex, more

dangerous machine better demonstrates that its makers are big, smart, powerful men.[xi]

When the novelty wears off, the technology enters a phase where it's only possible to make news when something goes wrong. The incentives that take over now are that the owners want to avoid that news, and they also want to drive down costs. Machines shrink, crew positions disappear, and whole panels of instruments are replaced with screens.

This life cycle evolves over and over. At the time of this writing, AI is in the novelty phase. The accident at TMI-2 happened during the awkward transition of nuclear energy from "novel" to "mundane."

Our culture rewards size and complexity with money, status, and power. That, in turn, is a product of the stories we tell ourselves. That's where we focus next.

---

xi Of course, not all of them are men. But most of them are, and we suspect that this particular cognitive bias has a gender affiliation.

HIDING BAD NEWS

Another habit that exacerbated the problem at Three Mile Island was the instinct to hide bad news. This is very rarely a conspiracy; rather, it is a natural cognitive bias. Most individuals are likely lying to themselves long before they are intentionally lying to anyone else.

At the 6 a.m. conference call with managers and executives, key information was omitted: The operators did not report that the PORV block valve had been open. We can't know whether the miscommunication was intentional, but the great majority of communication "errors" fall in the direction of downplaying a problem.

Initial statements from official spokespeople illustrate the impulse to minimize the problem. For example, one communications official told a local radio station at 8:25 a.m.:

> "There was a problem with a feedwater pump. The plant is shut down. We're working on it. There's no danger off-site. No danger to the general public."[2]

Why were they talking to a radio station at 8:25 a.m.? Funny story! A half hour earlier, a traffic reporter for a local radio station heard via a police scanner that firefighters had been called to the plant. He could also see, with his own eyes, that the cooling towers had stopped emitting steam. (Thus, he knew more about the current state of the plant than the operators did.) He called it in to his newsroom. The station's news director called the Three Mile Island switchboard. History does not record why, but he was connected directly to a phone in the control room. He heard sirens, chaos, and "I can't talk now; we've got a problem." Local media fired up immediately.

The government threw together a press conference. At 10:55 a.m., Lt. Gov. William Scranton III addressed the room:

> "Everything is under control. There is and was no danger to public health and safety. The incident occurred due to a

*malfunction in the turbine system. There was a small release of radiation to the environment. All safety equipment functioned properly."*³

In fact, neither the reactor plant nor the story were *remotely* under control. The press conference went off course immediately. Within the first minute, a nuclear engineer from the PA Department of Environmental Resources talked about fallout miles from the plant. Many of the remaining questions were irrelevant speculation about radiation in milk and technical details that the press corps did not understand. Similar press conferences and interviews multiplied over the next two days as the plant, the Pennsylvania government, and the feds all tried to "correct" one another's errors.

On the Friday two days after the accident, the governor recommended that pregnant women and preschool-age children evacuate to a distance of five miles.

After being assured for decades that a loss of control at a nuclear plant was impossible, the public was shocked into crisis conditions. Despite the urgent need to assemble a new story, there were no facts available except a mess of defensive, incoherent statements. The story that solidified was: We were lied to; the plant is incredibly dangerous; somebody screwed up; and *they are still lying now.*

An incredible coincidence confused the public understanding of the accident even more. Two weeks before the meltdown at TMI, the major motion picture *The China Syndrome* hit theaters. The plot of *The China Syndrome* is that a mechanical failure at a nuclear plant cascades into a core meltdown, after which the utility and the government conspire to cover it up. The control room and interior sets in the movie look identical to pictures of Three Mile Island. The movie contains dialogue that the meltdown "could render an area the size of Pennsylvania permanently uninhabitable." People have conflated images and press reports from the movie and the accident ever since.

The instinct to hide bad news months earlier, when an eerily similar incident took place at the Davis-Besse Nuclear Power Station near Toledo, Ohio, directly contributed to the accident at Three Mile Island.

THE "NONACCIDENT" AT DAVIS-BESSE

Davis-Besse had the same reactor model as Three Mile Island. On September 24, 1977, the reactor was operating at 9 percent of capacity. A control circuit malfunctioned and blocked water in the secondary loop. As the water boiled off, it tripped a safety system that sealed off the steam generator.

From here on, the events followed the same course as at TMI-2. Pressure increased until the PORV opened. It jammed open, but the control room indicator said "closed." The water level increased as it leaked out of the pressurizer. The operators scrammed the reactor. Since the reactor had been running at only 9 percent, the remaining decay heat was a negligible fifteen megawatts. The sequence was the same as Three Mile Island, but in slow motion. The ECCS turned on high-pressure injection, but the water level continued to increase. The Davis-Besse operators made the same decision as the Three Mile Island operators, shutting off high-pressure injection.

After about ten minutes, the supervisor realized that the reactor core was boiling. After twenty-two minutes, they hit on the correct solution: to close the leak in the broken PORV.

Mechanically, what saved Davis-Besse from being the namesake for nuclear danger, instead of Three Mile Island, was that the operators found the broken valve and blocked it in time. They had more time to think, and they also sacrificed the goal of "lower the water level in the pressurizer" sooner. As soon as the supervisor realized that the core was not acting as expected, they discounted the trained procedure and started looking for other explanations.

As a nondisaster, the incident at Davis-Besse triggered a noncrisis response inside its enclosing organization. Memos were sent from the plant management to the manufacturer. Memos were circulated among various engineering divisions. Much passive voice was used. The fact that the official training had caused the operators to shut down the HPI pumps, potentially causing a meltdown, was raised on

November 1, 1977. A handwritten response from a field rep came back nine days later, dismissing the idea and restating the ingrained and incorrect rationale for prioritizing the water level.

This argument slowly made its way up the ranks of the manufacturer. In February 1978, the safety-systems engineering manager issued a decision that the training and manuals needed to be changed. Six months after that, the field support staff responded that they had not implemented the change and would not do so. They remained committed to their decades-old, irrelevant Navy training.

The bureaucracy swallowed critical information: that the operators were unprepared for the failure of a PORV.

The reactor at Three Mile Island melted down seven months later.

SENSEMAKING AND CRISIS AT THREE MILE ISLAND

Let's apply our beliefs about sensemaking and crisis and see what fits.

Crisis Conditions Are Necessary for Rapid Change

We have no challenge at all defining the TMI accident as a "crisis." There was an obvious and urgent existential threat to the staff in the control room, the power company, and eventually the entire nuclear power industry—not to mention the town. The confusion that resulted when the temperature and pressure defied expectations is a textbook case of fundamental surprise. Lacking any mental model that could explain the data they were seeing, there was a breakdown in sensemaking. There was an inherent deadline (the impending meltdown), which the operators didn't exactly know, though the fast-unfolding events implied an urgency that was sufficient to motivate action.

The accident happened inside a system of interactions between the plant manufacturer, operator, and government. Each was a large and complex organization of its own. The macroscale behavior of the whole system also shows three of our five signs of crisis. The fundamental surprise was that, despite failure probabilities calculated to nine decimal places, the TMI-2 reactor melted down. The accident threatened to end the nuclear power industry (and it more or less did). Collective sensemaking disintegrated in the wider system, as each powerful organization chose a version of the story that placed most of the blame on someone else, and none of these stories lined up.

Sensemaking Is Motivated by a Sense of One's Own Identity

The operators' training—heavily influenced by Navy protocols—shaped their sense of reactions and priorities. On submarines, keeping the pressurizer from "going solid" was a matter of life and death, so operators were trained to avoid that scenario. When they saw increasing pressurizer levels at TMI, they followed that conditioning, even though this situation was different. Their actions aligned with their identity, even when reality demanded otherwise. Similarly, politicians confidently sought to assure the public and appear "in charge" well before they had even a basic grasp of the facts.

Sensemaking Is Retrospective

The narrative around Three Mile Island wasn't constructed in real time—it was assembled afterward, in congressional inquiries, media reports, and public imagination. The process of trying to make sense of the event led to conflicting yet durable stories: some emphasizing deception and cover-up, others focusing on human error and system complexity. These retrospective stories became hard to dislodge because they were internally consistent and emotionally resonant.

Sensemaking Is an Active Process of Cocreating Your Environment

This is also easy to verify. The actions taken by the operators created the situation that the operators then had to contend with, every step of the way. The mishap with the clogged filter tripped the feedwater pumps, then the turbine, then the reactor scram, then the PORV, then the high-pressure injection. The operators turned off HPI, causing the loss of coolant. Finally, the operators blocked the PORV, which stopped the bleeding and halted the meltdown with "only" part of the core destroyed. Each of these actions was a result of the operators choosing to focus on some facts at the expense of others.

Later, the managers who tried to take control of the public narrative created more conditions with which they would have to reckon. Comments to the effect of "Radiation release is no big deal" and "We don't have to explain ourselves to you people anyway" destroyed credibility.[xii] Nobody from a government agency had much of a chance, because they did not have any direct access to information on the ground in the first place. When the governor's office recommended that pregnant women and small children evacuate, the result was that *everyone*, not just those two groups, rushed to get out of town. Some observers described the evacuation as panicky and chaotic, while others insist it was orderly; probably all can agree that it was stressful, and put strain on resources such as police and highways.

---

xii This was a vice president of the company that operated the reactor (Met-Ed) speaking to the press. The same vice president later explained that he ignored the PR staff's advice because "PR isn't a real field. It's not like engineering. Anyone can do it." This does seem to have been their belief: Met-Ed went on to promote a security guard to its communications office six months later.

Sensemaking Is Social

Some of the strategies that we now advocate for navigating a crisis were taken for granted in 1979. All the operators were present together in a control room, surrounded by all the instrumentation that the plant had to offer, because there was no other way to do it back then. They were able to identify a pattern of readings that suggested the pressurizer was becoming overfilled, and agree on the course of actions to correct it. No less than three operators arrived at this shared reality in less than five minutes. In our experience with training exercises and real life, we rarely see this kind of reaction time. In trials run over a phone or video conference, we sometimes don't see any shared reality emerge at all, *ever*, which is part of why we are so against remote crisis engineering efforts.

Of course, the immediate reaction of the operators was wrong for the circumstances. Unfortunately for them and for TMI, they did not improvise the correct solution like the crew at Davis-Besse. The critical question about the PORV wasn't asked or answered until two hours later, when there were more people in the control room and even more on a conference call. The testimony and accident records do not say much about this period, but based on our experience, we'd expect that the main activity was explaining and re-explaining the situation to each new person who turned up; there was also a highly repetitive debate about the actions already taken and possible solutions. In any event, quite a few people became involved before the solution was discovered (too late).

Sensemaking Is Continuous, with No Start or End

The process of sensemaking had no clear beginning and no clear ending. When trying to write the story of the accident, where do we start? Official accounts begin with the shutdown of the turbine generator, which is unexplained. With hindsight, we know that the turbine tripped because the feedwater was cut off, which was because valves failed, which was because of a problem flushing a water filter.[xiii] That's where we began in the short summary in this book, but that, too, is arbitrary. Why were the resins in the water filters so troublesome? Why did the technician try to clear a clogged filter with compressed air? What else did they try?

If for some reason we wanted to blame the entire accident on the failure of the filter, we could write the story to do so. Or, if we want to blame the operators, we can begin the story at the point where the automatic failsafes have engaged and the core is stable until the operators turn off the high-pressure injection. This is how the official inquisition and the congressional oversight process chose to frame it.

Likewise, where does the TMI story *end*? The last episodes to get wide attention were the congressional hearings and the release of the Presidential Commission report. The physical state of the reactor core still wasn't even known at that time, let alone any consequences for the plant site or the surrounding area. All anyone could agree on was that it was the worst accident in the history of nuclear power.[xiv]

The industry, regulators, and the wider government all set out to blame *somebody* besides themselves, and assure the public that everything was under control. The focus became the unlucky operators and managers at the controls that day. The picture drawn by the hearings was that power plants were run by lazy, incompetent utility companies who didn't care about the consequences of their decisions and would lie to cover them up.

The Pennsylvania House of Representatives maintains a collection of TMI paraphernalia, which your authors visited. The local population was surprisingly sanguine in the aftermath of the accident. A bumper sticker announced, "Hell, No! I Don't Glow!" Visitors could buy official TMI radiation, allegedly canned on location in

---

xiii In case you read any of the more technical documentation, they are called "polishers" in the industry jargon.

xiv (Homer Simpson voice:) "Worst accident...*so far*." Chernobyl was seven years in the future.

Mikey Dickerson, 2024

Middletown. These don't seem to be the products of genuinely terrified people. Perhaps this is because this is an area with coal mining and other heavy industries, which are all dangerous. For all the fuss, no one was actually injured in the vicinity of the TMI accident.

One could end the story there. But the reverberations of the accident did not become clear for decades. Interest groups that were already fiercely opposed to nuclear weapons joined up with the environmental movement, creating a coalition that fought hard against any form of the technology, deliberately conflating power stations and bombs to maximize fear of both. They were so successful that it would be forty-four years before the next reactor was successfully brought online in the United States.[xv]

This is not the last word, either. Two generations later, the abandonment of nuclear power is beginning to be seen as a policy failure, since we made up for it with coal and gas.

---

xv This was Vogtle 3 in Waynesboro, Georgia, which entered operation on July 31, 2023.

Sensemaking Is Driven by Plausibility More Than Accuracy

Eventually, when the events of the accident were better understood, the consensus among physicists was that the amount of radiation that escaped into the environment was very small, amounting to less than one year's worth of normal background radiation. Epidemiologists searched for long-term effects, starting with a 1990 study from Columbia, and continue to the present day.[xvi] The published data show approximately a 0.034 percent increase across all types of cancer within two years and ten miles of the accident.[4]

There is no agreement on what that means. The authors of the study consider there to be no connection between the accident and the cancer rate. A reanalysis paid for by lawyers for the residents did not change the numbers, but argued that the 0.034 percent increase was significant.[5] It also included some novel genetic tests that seemed concerning. Numerous other people weighed in, pointing out facts such as: Methods for cancer screening have changed and generally gotten more sensitive; this area of Pennsylvania has much higher than the average amount of radon; and two nearby counties that do not contain nuclear plants have even higher cancer differentials under the same methodology.

No one is convincing anyone else.

Back in 1979, here is one plausible story that could have been assembled with the available facts:

> *There was a serious accident at the TMI plant. It caused us to be exposed to radiation. We don't know how much. The people that know are lying about it, and are afraid of the consequences.*

With that framework, it is no trouble to fit in the subsequent

---

[xvi] One of our colleagues from Google was a child in Middletown, Pennsylvania, at the time of the accident. He still receives annual health questionnaires from the Pennsylvania Department of Public Health.

facts: The published dose measurements all came from a combination of regulators and plant officials, and they therefore can't be trusted. The epidemiology studies are also government funded. Any particular resident likely has other anecdotal facts to toss on top: Perhaps they know one or two people who developed cancer in the years after 1979. Those people didn't have cancer before!

Another plausible story is no less supportable:

> *There was a serious accident at the TMI plant. It was caused by a mishmash of mechanical failures and honest mistakes. There was confusion and disarray after the accident because the experts didn't understand any more than we did. But decades of careful monitoring since then have confirmed that there was barely any radiation released in the accident, and there were no effects on humans or the natural environment outside the plant itself.*

Sensemaking suggests that both of these stories have enough of a factual framework that no further facts will dislodge either one. Our prediction as crisis engineers is that there will be no significant movement toward any consensus unless another major shock precipitates a crisis in the nuclear industry.

We may live to see such a shock, because the story of Three Mile Island continues today. Microsoft recently signed an agreement that will bring TMI-1 back online for another twenty years of operation. They intend to use the plant to power AI data centers. AI, of course, needs so much power because it is in the novelty phase, where bigger is better and safety and usefulness are afterthoughts. Time is a flat circle.

We chose a detailed look at TMI because it's hard to imagine a better illustration of why we believe sensemaking is the key to both tactical and strategic crisis engineering. We can't fit them all in the book, but there are several other well-studied crises with similar properties: the *Challenger* and *Columbia* space shuttle accidents, the

Therac-25 radiotherapy machine, Chernobyl, and the Mann Gulch wildfire. In all cases, relatively small alterations to the crisis management effort could have restored sensemaking and averted disaster.

Our next detailed case study—HealthCare.gov—is also a crisis told through the lens of sensemaking, but this time, there were crisis engineers on the case. We can narrate this one firsthand, because we were there.

1.3

HealthCare.gov

> They start saying to themselves, as if it had never occurred to them before, "What the hell does the government know about making a parachute that will open five seconds after it starts to fall? Not a damn thing. They just farm it out to some fly-by-night outfit that makes the lowest bid."
>
> —NORMAN MACLEAN, *YOUNG MEN AND FIRE*

The closest thing to a Three Mile Island–scale accident yet to befall the U.S. government tech ecosystem is the failed launch of the HealthCare.gov health insurance marketplace website in 2013. Most Americans were unable to even open the website, let alone shop and apply for insurance. The crisis fundamentally threatened the viability of the new Affordable Care Act legislation that made affordable health insurance available to every American for the first time. HealthCare.gov was the primary portal through which millions were expected to enroll in coverage, making its failure a direct threat to the law's core promise. News media covered the failure 24/7, often demonstrating the site's failure to load on live television.

Years later, it still casts a shadow over an entire sector of government services. "We don't want to be the next HealthCare.gov," people in any agency will say before attempting anything that might involve a computer.

We consider ourselves fortunate, and humbled, to have been there to resolve this crisis. HealthCare.gov was the most difficult case of crisis engineering we have done.[i] We've done our best to deconstruct the experience now—because sensemaking is retrospective!—and see how well it fits with the lessons we took from other disasters.

THE RESCUE OF HEALTHCARE.GOV

In mid-October 2013, Mikey agreed to do a small favor for a friend from the Obama presidential campaign. This favor, as best he could tell, involved joining a chaotic conference call, for which he had no context. On the call, Todd Park, the chief technology officer of the United States, introduced Mikey and a handful of other technologists to the leaders of several government agencies. The call quickly escalated into a plea for these technologists to come help save HealthCare.gov.

None of the government officials on the call had any role in implementing the website. That kind of technical work in government is done by contractors.[ii] When you visited any technical staff on the government side, they would pull out a massive full-color "architecture diagram." The diagram was several years old, with boxes labeled

---

[i] (Homer Simpson voice:) "Most difficult one *so far*!"
[ii] The word "contractor" may trip up readers from the tech industry. The only version of contracting that most Big Tech workers have seen is staff augmentation, where a company like Google will occasionally hire "contractors." In that context, the term refers to individual people who are added to existing teams, often with the same responsibilities as other Google employees. The key differences are the colors of their badges and whether they can access certain perks like free food and trips to Disney. This is not what the term means in government; when someone says "contractor" they mean the entire *company* was awarded a contract. When a news report says "Fifty-five contractors worked on HealthCare.gov," West Coast readers get the wrong idea. That isn't fifty-five people—it's fifty-five *companies*. The number of actual people involved is hazy, and only gets hazier the more you ask.

with abstract ideas like "Application Server" or "Data Storage." If you don't know what these are, that's fine—neither did the officials who made the diagrams. There was no point asking them detailed questions—those could only be answered by the contractors.

Soon, five people from that original phone call—Greg Gershman, Paul Smith, Gabriel Burt, Ryan Panchadsaram, and Mikey—found themselves piling out of a White House car and into the Herndon, Virginia, office of one of the largest contractors, CGI Federal. Our instructions were to determine whether there was any hope of making HealthCare.gov work. We had three days.

We believed we had been summoned to a five-alarm computer fire, and were immediately baffled by the lack of activity. It was now three weeks after the website's failed launch. The parking lot was half full. The ocean of nondescript gray cubicles was quiet. Peeking at computer screens, we saw email, Jira tickets, and Outlook calendar entries for routine meetings. What we did not see was HealthCare.gov, let alone the news. Coffee mugs were filled and sipped at their regular *Office Space*-esque pace.

For the first day or two, we tried to find basic facts about the website. We split up and approached random cubicles, asking each person to introduce us to the next person we should talk to. We turned an empty office into our home base (quickly dubbed the "war room"), where we compared notes and constructed what maps of reality we could on the whiteboard. Most of the work to make sense of the situation actually happened over lunch or dinner. The team would assemble at a nearby Italian restaurant, where TVs over the bar blared the endless CNN coverage of HealthCare.gov.

Our main discovery was that no one was coordinating the fifty-five contractors, nor was anyone attempting to debug problems end to end.

So, we began to do this ourselves. By the second day, Greg and Paul identified and fixed a small misconfiguration. As a result, the metric for "response time" (how fast the website loads) dropped from eight seconds to one second. Progress like this convinced us that there were lots of fixable problems. At the end of the three days, we gave our best guess: The site could be made to work before the end of the calendar year.

The plan to stay "a few days" and evaluate the situation turned into an open-ended commitment to fix an unknown number of problems of unknown size. This was, at first, confusing. There were literally thousands of people whose actual job involved this website. Why did they need *us* to run the response effort?

Some people encouraged us to think that we had rare and valuable skills that were nowhere to be found in the contractor ecosystem. There is a little truth to this, but there is a different explanation that is more true: The incredibly complex contract structures behind HealthCare.gov were a masterful choreography of checks and balances, all working together to ensure that no one person or company was accountable for the whole project actually working. That structure protected everyone from experiencing any consequences, as long as they did not attract any attention to themselves.

In that moment, for better or worse, an outsider could take control of the ship—and so that's what we did.

We settled into a routine in the war room. Twice a day, the incident lead (Mikey) attempted to identify the worst issue affecting the site at that instant and try some action to fix it. These mundane practices—designating a room where things happened, and making actions happen at least twice a day—were a crucial part of how the website would eventually work well enough to get millions of customers health coverage. We call this a "crisis engineering center" in the toolkit section later.

Another critical feature of the war room was a shared view of system performance, which attempted to capture the end-to-end experience of a user on the website. Imagine if Three Mile Island had a display showing an end-to-end picture of water flowing through the entire system; this was similar.

The early weeks of the HealthCare.gov rescue mainly focused on fixing any small targets of opportunity that we stumbled upon, combined with relentless daily pressure on all of the contractors to install monitoring on their individual components so we could start to see the entire system in one shared view.

Outside of the war room, Jeff Zients, then an advisor to President Obama, took over the job of running the daily press conferences. His apparent strategy was to dull interest from the press by overwhelming them with exhaustive detail. Jeff and Todd became regular visitors to the war room. The team did our best to explain the up-to-the-minute status, and a few hours later, Jeff and his team did *their* best to convert those details into something the press could consume.

The war room itself went through two changes. Our group outgrew the size of the original conference room within the first week. When we discovered that the entire vacant floor above us was already leased by CGI Federal, we arranged rows of tables and two TVs in it and declared it the new war room—operating on the assumption that under these crisis conditions, we could sort out any official approvals later. By now, most of the contractors had sent permanent representatives on-site, and they assumed positions at the tables, too. Pictures from this period show how it was now possible for people from many different companies to gather around the same TV to look at the same evidence at the same time, which sped up the pace of troubleshooting considerably.

This was the most convenient and spacious of the war rooms, but it didn't last long. Someone eventually started asking why we weren't using the exchange operations center (XOC), a war room that had already been bought and paid for well in advance of the launch, but that no one involved in the launch seemed to know about.[iii]

Before long, another official White House transport was dropping us off at a mall in Columbia, Maryland, where to our wonderment, we found a whole facility purpose-built as a HealthCare.gov war room. It had not two but *twelve* TVs. It had a half dozen multi-person desks with monitors that we could make to dramatically rise out of the table when things got serious. The building was smaller,

---

[iii] "The exchange" was one of the government nicknames for HealthCare.gov, along with "federally facilitated marketplace," or FFM.

Matthew Weaver in the XOC
Mikey Dickerson

harder to access, and had fewer bathrooms than the previous one. But it looked the part, and it was closer to the headquarters of the government agency ostensibly in charge of HealthCare.gov: the Centers for Medicare and Medicaid Services (CMS). This final war room was hereafter known as "the XOC."

The nearby mall had a food court, at least.

A consequence of the move to the XOC was that it broke up the original crisis team. Mikey left the team and moved to the XOC, where he continued serving as incident lead twice a day. A routine developed. When the website experienced an outage, real-time group troubleshooting took priority over all else. The rest of the time, ideas for possible improvements were entertained from anyone in the crowd.

Elsewhere, the website developers, under the watchful eye of the rest of the crisis team, continued to work on code. Approvals to make changes to the site came from the "change control board," a government committee whose membership and decision criteria was something of a mystery. It held a daily phone call, but you couldn't see or know who had dialed in. Someone recited the list of proposed changes, and if nobody spoke up with an objection, *all changes were approved by default*. During the rescue effort, this call happened every afternoon (as opposed to once a week), with changes deployed every night. However, the XOC had special permission to deploy emergency changes outside of this process if necessary.

The job of XOC incident lead[iv] was too exhausting for Mikey to keep doing alone for very long. The government and some of the contractors worked together to make it possible to quickly hire a few more people. (Outside of the crisis, this would have taken months, at best.) Within a few days, Mikey was able to bring in more incident leads to work in shifts, so they all had a chance of sleeping through the night.

To the extent the "rescue" had an end, it was the day we called Vast Majority Sunday.

Jeff Zients, still managing the daily press conference, told the press to expect that by the first Sunday in December, the "vast majority" of people would be able to enroll in health insurance. This was an attempt to synthesize a publicly accessible story out of the daily chaos at the XOC. It was as reasonable a guess as anyone could have made, but we had no way of seeing that far into the future to know for sure.

As Vast Majority Sunday approached, the site was having good days and bad days. On a good day, servers did not crash, error rates were stable, and we could count tens of thousands of successfully filed applications. On a bad day, response times oscillated wildly, pages stalled, and errors piled up. Vast Majority Sunday itself started out looking like a bad day. It all came down to lagging performance in the primary document database.[v]

The database contractor's engineers speculated that the problem

---

[iv] The job would later be named "pit boss," in honor of the XOC's resemblance to casinos with table games.

[v] On this type of system, it is usually the case that the database appears to be the worst performance bottleneck. It is because the database is doing most of the hard work, so it is the first place overload becomes visible. Years of gossip notwithstanding, it is not likely that anything would have been different with a different database.

was in the underlying operating system. In desperation, we invoked our emergency authority to deploy a midmorning change. Calm returned. CNN's attempts to load the site on live TV did not show errors. Vast Majority Sunday was declared a success, and the cable news media moved on.

Notably, nothing particularly changed on that day with respect to the actual metrics we were tracking on the dashboard, but it was an emotional turning point. Mikey remembers being able to sit down while on duty for the first time.

While the news vans dispersed, the XOC did not shut down that day. By January, the crisis team had grown to a dozen or so people. This was when Weaver joined.

With the implementation of "incident management" conventions (such as keeping a running list of actions taken, with time stamps), the routine became predictable to the point that the incident lead job could be interchangeable. Most of the new incident leads stayed around two to four weeks and went home. Mikey went home in March, returning several more times until the enrollment window finally ended in May 2014.

In the following years, the XOC team became a function of the United States Digital Service (USDS), an office inside the White House that was created after the HealthCare.gov debacle. Mikey led USDS from August 2014 until the end of the Obama administration in 2017.[vi] The XOC war room continued to exist, and vestiges of the daily routines continued to run, for ten more years.

WHY WAS THE LAUNCH UNSUCCESSFUL?

As at Three Mile Island, the technical issues giving HealthCare.gov trouble were things that had already been solved elsewhere in the industry—in this case, widely. Critically, the site operators lacked any venue for effective sensemaking. Creativity and improvisation

---

vi President Donald Trump effectively disbanded the USDS by renaming it the U.S. DOGE Service and turning it over to Elon Musk in 2025.

mostly happened in the higher levels of the controlling bureaucracy, forcing the federal government to develop radical new behaviors under time pressure. We will examine each of these angles and the way they shaped our idea of crisis engineering.

Machine Failures

Mechanical failures like a jammed valve are rarely a factor in a modern computer system. Sure, wires can corrode and spinning disks can crash, but cloud providers attempt to hide the frailties of the underlying hardware. For most sites, this works. Even the HealthCare.gov physical hardware plant caused very few issues that we discovered.

Unfortunately, the simplification gained by modernization has been more than offset by an explosion in software complexity. Bugs and misconfigurations among virtual machines have taken the place of "machine" failures. Of these, HealthCare.gov had too many to count, though there were a few common patterns:

SYSTEM INTEGRATION FALLS THROUGH THE CRACKS
Dozens of problems were in basic system integration, like app servers with the wrong name for their cache server. Inside the maze of contracts and contractors, these issues seemed to everyone like someone else's problem.

OVERCOMPLICATION CAUSES INSTABILITY
Each system component had dozens of configuration options whose behavior in the overall operating environment was hard to predict. For instance, each of the load balancers (something that spreads out work evenly across computers) tended to be configured in the cleverest and most complicated way. Whether this is good or bad is hypersensitive to local conditions—but in a complex system, cleverness is almost always bad. It fails to solve problems unless the clever programmer has guessed precisely what the problem is going to be, which they can't. You pay a huge price by making the behavior unpredictable to operators.

Across many iterations, we simplified components like load balancers, and removed entire tiers of complexity, which calmed down the worst of the wild oscillations.

ABSTRACTIONS LEAK

Another class of problem that was very hard to debug was a bad interaction between layers of abstraction that weren't aware of one another. This can cause new behavior to emerge that isn't visible from either layer on its own. (Recall that this was how we defined "complex system.")

It's likely that you've encountered this type of bug in places where big corporations meet. A petty example from one of our lives is broken credit card benefits. American Express believes it has enrolled Mikey in the status program at Hilton hotels. Hilton does not agree with this fact. There is no mistake visible in either company's computer, so there seems to be no power on Earth that can get Mikey his free bottle of water.

NERDY SIDEBAR

An actual example: In the midst of the debugging effort of HealthCare.gov, the Linux kernel changed behavior to enable a feature called "transparent huge pages," which means it manages memory in larger blocks. On some workloads, this improves performance. On the primary storage for HealthCare.gov documents, it was catastrophic. We don't know whether it was a bad interaction with VMware virtualization, or with something fancy MarkLogic does with memory allocation, or both. After a lot of time spent trying to isolate the cause among the many rapid changes to the site, we eventually disabled the kernel feature. This pathology almost caused the Battle of Vast Majority Sunday to be lost.

Operator Errors

At Three Mile Island, there were humans sitting at desks pushing buttons and pulling levers to manage the daily production of power. Websites are not run like this. The staff in the XOC, and their direct managers, were approximately the "operators" of HealthCare.gov.

In this story, we also have to look for patterns across longer time periods, because the HealthCare.gov crisis was closer to eight months than eight minutes.

INEFFECTIVE TROUBLESHOOTING

Complex systems generate problems that seem as if they are intentionally designed to escape your troubleshooting tools. For instance, as SREs at Google, we learned to search the record for clues at the time an outage began, and if that fails, to roll back recent changes.

HealthCare.gov bugs often defeated both these strategies. One particularly difficult bug stands out as an example.

It started in November with "brownouts"—periods of overall slow performance. No obvious triggers happened at the beginning of each one. After several days, all we learned was that the servers seemed to be waiting for the database. If you restarted any particular slow server, it would snap back to normal performance. But the overall situation got steadily worse. Soon the brownouts started midmorning and continued through the rest of the day.

We reversed all recent changes. No improvement. Restarting the slow servers, which was like bailing water out of a leaking boat, couldn't keep up. The behavior was also erratic: You could coast for hours and then suddenly everything went crazy again.

Because it looked like database slowness, the database engineers wanted to figure it out. One of them called Mikey over late at night. The engineer had discovered that some of the files representing applications were absurdly large. A normal file was a few kilobytes, but these were tens of megabytes—more than a thousand times bigger.

The obvious next step was to look at one file and see what happened. But here was trouble: This was live data, so this file contained a real person's health care application. It probably contained sensitive health information regulated by HIPAA. We were not authorized to look at it.

Instead, we used the crude tools that we had available in production to look at the *structure* of the file, without its actual data. We immediately saw the problem: duplication in the "dependents" data structure. ("Dependents" meaning your health care dependents: kids, typically.) We thought long and hard, but could not imagine any legitimate situation where an individual person had a hundred thousand dependents.

Given this clue, others figured out the underlying cause: When a user performed certain actions, like clicking the "back" button on the browser, the app created two copies of the dependents list. The next time they clicked back, they'd have four copies, and so on.

For reasons we will never know, 112 users of HealthCare.gov habitually triggered this bug. Every time they doubled the size of their application, the time required to fetch it from the database doubled. Some of them had persevered to the point where every time they clicked "Next," it took the website forty-five minutes to load.

The contractor fixed the software bug immediately, but the problem remained, because the 112 giant files were still there. We expected any formal proposal to fix the situation would lead to weeks of decision gridlock over what we should, or could, do to these 112 applications. In the meantime, no one could use the site. Thus, at another late-night conference in the war room, the 112 unparseable applications were quietly deleted.[vii] Site performance stabilized.

[vii] The next time those 112 people logged on, they would have to start over from scratch. This was not great, but it seemed reasonable that they would persist in completing the application, considering that yesterday they were willing to wait forty-five minutes per page load. Any other solution would have amounted to modifying their application without their knowledge, which seemed much worse.

OVERLEARNED LESSONS

Paradoxically, several times, we saw the organization try too hard to capture lessons learned.

Deploying code changes every day meant that every night, the site had to be fully shut down and brought back up.[viii] One component was particularly fragile, and would crash when too much traffic hit the site too quickly. The mitigation was to open the doors slowly: For the first hour, only a trickle of visitors were allowed to access the site. When Mikey returned to the XOC a year later, improvements had eliminated the problematic component, so there was no need for the trickle. But this one-hour warm-up period continued, now enshrined in a "method of procedure." Every maintenance window was now longer than it needed to be, trickling traffic for the first hour for no reason.

It was a learning experience for Mikey when he discovered this practice had not only survived well past its usefulness, but that he was unable to get rid of it. Explaining that he was the one who created the practice in the first place, for reasons that no longer exist, made no difference. The crisis was over, and the priority was once again to preserve the status quo.

Organizational Failures

From the perspective of someone who wants to prevent their next program from turning into a HealthCare.gov, the organizational mistakes are what matter. The procurement and architecture decisions made years prior virtually guaranteed the deficiencies we found on day one.

DIFFUSE RESPONSIBILITIES

The HealthCare.gov project was structured from the start to create a multiplicity of "small" contracts pieced out to dozens of firms.[ix]

---

viii Of course there are techniques for avoiding this, but HealthCare.gov did not have them.

ix Don't forget this is "government small": each contract was still tens of millions of dollars.

Some contracts were meant to encapsulate a box on the architecture diagram, but others were assigned an entire horizontal layer of the tech stack. One contractor was responsible for the hardware, another for the virtual machines, a third for the operating systems, a fourth for the software packages, and so on.

Since application components run atop all of the above, there was a matrix of responsibility that caused any specific problem to land in the jurisdiction of several contractors at once. Before the XOC, each contractor worked in isolation. The government staff at CMS were the only ones with theoretical visibility across the systems, but again, they had no technical experience.

CMS thus gave itself the job of coordinating the work of fifty-five contractors, using the part-time efforts of a few government officials. We don't think anyone on Earth could have been successful in that role, and CMS had two disadvantages making success even less likely:

LACK OF EXPERIENCE

Until the Affordable Care Act, CMS concerned itself primarily with running Medicare and Medicaid. (Recall that the M in CMS stands for "Medicare and Medicaid.") These programs consist of processing claims submitted by doctors' offices and sending out payments. They can be done in batch jobs. Mistakes can be corrected at an end-of-quarter "true-up." If the processing jams, CMS can send advance payments and figure the details out later.

This is quite unlike running a transactional website. HealthCare.gov could have millions of customers at once, and the system needs to respond in milliseconds to each one. They can't be put into a queue and processed one at a time at a comfortable pace. If there are mistakes, the user sees them. This is much harder to fix than a back-office transaction with a long-standing customer.

There is specialized expertise, technology, and folklore that allows sites like Google and Amazon to run at utility-level reliability. At the time HealthCare.gov launched, Google had well over a thousand people dedicated to site reliability engineering, a competitive and high-paying field. In contrast, CMS had never run a transactional website before, so it had no such people around, and didn't know what it didn't know.

LACK OF URGENCY

The environment we encountered arriving on the scene in October 2013 was eerily serene. Business was conducted over email, and emails were answered during business hours, at the ordinary stately pace. Political appointees, seeing clips of themselves on CNN, were apoplectic, but none of that energy escaped the executive suites.

Henry Chao, a deputy chief information officer (CIO) at CMS, disputes this in his book *Success or Failure?* In his telling, the CMS team had been working itself into the ground for years ahead of the launch.

We don't know what the process was like before the launch of HealthCare.gov, because we weren't there. But if Henry's depiction is correct, it's hardly less of an indictment of the organization. It says that CMS had already experienced the crisis, which had already burned out and resolved into a rationalized failure, before the site even launched. It had months or years to adapt and didn't do it.

Cultural Failures

Looking at Three Mile Island, we picked out three features of the culture that raised the risk of a bad accident:

- Fetish for complexity
- Hiding bad news
- Inability to learn from surprising events

There is ample evidence of the same factors inside HealthCare.gov.

While the site was deficient in some areas (like having no monitoring to show if the site was up, or slow), it was massively overprovisioned in others. The data center was big enough to run a site ten times its size. Caches and load balancers were bolted onto nearly every serving component, creating operational issues like the warm-up delay mentioned earlier. This is not to mention the legislation itself, which was so complicated that applying for coverage could require upward of forty separate forms.[x] Top to bottom, people were responding to the cultural bias that says smart things are complicated, and complicated things are smart.

As for "hiding bad news," where do we even start? According to their congressional testimony, several White House leaders asked CMS for updates throughout the implementation phase, and always received versions of "everything's going fine."

When it comes to not learning from surprising events, there may even be an analog here to the Davis-Besse incident, which foreshadowed the TMI accident. Many of the senior CMS staff who pitched in *after* the HealthCare.gov launch were veterans of the 2006 launch of Medicare Part D. According to institutional memory, that was also rocky at first, but then "smoothed out."[1] Unfortunately, not many details of how exactly it was smoothed out survived. If anything, the memory of Part D seemed to give some officials false confidence that if they just waited long enough, HealthCare.gov's problems would somehow all go away, too.

Besides those three widespread cultural features, government culture has specific issues of its own. It is a culture where there is no possibility of being rewarded for good performance, no possibility of consequences for (most) failures, and where the only thing that *can* lead to punishment is to say or do something *different* from the rest of the agency. Government culture focuses a lot on adherence to rules and procedures, and surprisingly little on what outcomes (if any) that process achieves. It's hard for outsiders to believe that with so many signs of trouble, nobody said anything, but they underestimate the

---

[x] The American problem with excessively complicated policy is discussed at length in books such as *Administrative Burden* by Pamela Herd and Don Moynihan, *Recoding America* by Jen Pahlka, and *Abundance* by Ezra Klein and Derek Thompson.

effect of spending one's whole career in such a stunted environment. Bureaucrats learn not to say anything in any forum that might raise alarms, while regularly saying things in side channels that will later prove they had done all they could.

Finally, coordinating contractors through a government program office is extra hard because contractor-government communication is stilted and formal. Lawsuits between contractors and agencies are routine, and the processes for making decisions and giving direction have evolved as if the actual goal is winning a future lawsuit, not delivering a service. Since lawsuits are common and feel inevitable, everyone behaves as though they're practicing for the witness stand from day one, intending to show how they adhered to every rule and said (or didn't say) all the right things at the right times. Face-to-face meetings are restricted by baroque rules about who can attend from each side. Lawyers are usually in the room. Agencies and their contractors act like divorced parents from the day their relationship begins.

SENSEMAKING AND CRISIS AT HEALTHCARE.GOV

HealthCare.gov was a formative experience for us and our ideas about crises. It inspired our interest in sensemaking. The following are a couple of additional examples of how sensemaking was necessary for resolving this crisis.

Crisis Conditions Are Necessary for Rapid Change

The pace of change in the government that happened on HealthCare.gov has become the benchmark against which other projects are compared, and a guiding example of the type of change possible as a result of a crisis. If it hadn't happened here, few people alive today would believe it was possible.

While the members of the crisis team, contractors, and frontline government supervisors were working sixteen-hour days in the XOC, their managers and executives were working similar hours in rooms nearby. Each of the two following feats would earn a Bronze Star in bureaucracy, if there were such a thing—and this is by no means an exhaustive list of the resulting changes.

CONTRACTS WERE RESTRUCTURED

As discussed, the original structure had each of the fifty-five contractors answerable to the government directly. Any coordination between two companies formally required playing telephone through the government. It's unsurprising that on opening day, there were an uncountable number of mistakes in the plumbing.

A few days in, it was obvious that a more effective plan was to have a central coordinator (the incident lead and twice-daily meetings) and to give contractors the ability to interact directly (the war room). CMS officials issued "technical direction letters" that brought contractor representatives on-site in a matter of days. Soon after, a new "systems integrator" contract was created, delegating one company the power to give technical direction to others. This company took on the job of hiring the people who became the crisis team.

If we had been asked in our subsequent White House jobs to install a new hierarchy in a preexisting web of fifty-five contractors that had already been struggling for years, we would have declined. It requires finding new money, designing new contract requirements, awarding new contracts, winning all the appeals and protests, modifying the preexisting contracts with new technical direction, and surviving *those* appeals and protests. An experienced administrator should expect any one of those steps to take six months or more, and they must happen serially. At HealthCare.gov, it was all done in a month.

CMS executives worked long hours, took risks far outside their comfort zones, and will never be recognized for it.

HIRING WAS ACCELERATED

No less astonishing was the fact that the crisis team grew to include dozens of people in the span of two months. Even more impressive, the very first person Mikey brought over from Google was a Canadian citizen, requiring a novel hiring process. All the new hires were processed as subcontractors earning the same hourly rate of $200, meaning that compensation and salary processes were bypassed as well.

This brings up another problem: There is often no way to reconcile private and public pay scales. The $200 number we just mentioned might not have raised eyebrows among our corporate readers. But government and civic tech folks are likely agog. This problem, unsolvable under noncrisis circumstances, was brushed aside during HealthCare.gov because *both* the government *and* the outside talent took the attitude that the price was unimportant. The pay gap for in-demand technical specialists is so large that neither side can get around it alone.

You don't have to take our word for it. Consider the following description of an April 2013 meeting from Deputy CIO Henry Chao:[2]

> *In one of these twelve-plus-hour briefings held in April of 2013, we were joined by several McKinsey consultants for discussion and to get a background for an assessment they were brought in to perform...I took the opportunity to ask Todd [Park] to advocate on behalf of my team to bring in four to six high-caliber technology people that not only understood technology, but how to work with the business teams to lead efforts to accelerate the process for prioritizing and defining requirements and translating them into building the [minimum viable product]. Somehow, even after dedicating a dozen hours going over all the issues and making a plea to bring in outside assistance, the request went unfulfilled.*

This is, of course, the same Todd who *did* bring in five technology people in October, for the effort that would get labeled "the rescue." Somewhere, there were immovable obstacles to running the same play in April.

Read any recent book on government and technology, and the top two candidates for "worst dysfunction" are procurement and hiring. Yet, a crisis created the opportunity to practically disregard them for a few months in the winter of 2013.

Sensemaking Is Social

There is no hope of a group consensus about what's going on if they all aren't consuming the same facts. This is why monitoring (or "observability") is crucial. One of the first things the crisis team did was use an existing tool (New Relic) to observe basic facts about the entire distributed system, such as whether the website was up, and how long it took for a page to load. If the rescue effort introduced any revolutionary technical change, this was it.

Most of the time, New Relic collected less complete and less accurate metrics than the artisanal monitoring consoles on each individual software component in play. But each one of those was different, restricted access to a different small subset of people, and used different language to describe the same things. Whatever New Relic lost in detail was more than made up by the ability to be understood by all.

Additionally, social hierarchy significantly influenced which ideas and theories got attention. The crisis team spent a lot of days, nights, weekends, and holidays together in a small room. There were popular kids, nerds, and the kids at the back of the bus. When there was a mystery available for anyone to solve, the desire for approval from the popular kids was as much of a motivator as anything else.

Schoolroom dynamics aside, there was a strong problem-solving capability among the crisis team that wasn't entirely explained by surface interactions. In recent years, managers have imagined in-person work in open-plan offices to be the best way to promote innovation and collaboration. The general assumption is that everyone has something to contribute to everyone else, and open-plan offices maximize chance encounters.

But this doesn't match what we saw in the XOC or in our past open-plan offices. In most group discussions, the same handful of people do almost all the talking. Truly random conversations that aren't simply an extension of a preexisting relationship are rare. Yet, if you remove all the passive and silent participants from the room, the consensus-building process fails. Everyone needs to feel like they had a role in the process, even if they didn't speak up. Humans in meetings react to confidence, tone, posture, and especially, watching the *other* watchers to learn who they should pay attention to.

We didn't have the words yet, but what we were observing was group sensemaking. The XOC was a perfect laboratory to watch how it works (and doesn't work).

Plausibility Beats Accuracy

We described how the perceived success of the rescue effort came down to Vast Majority Sunday. Jeff Zients and the communications team took a gamble when they chose that date weeks in advance. It wasn't driven by any engineering timeline; we refused to give any time estimates on the grounds that we didn't know how much needed to be fixed. While this was true, it was very frustrating for them to hear, and they marched ahead to declare a date without us.

Having announced the plan and the end date, the communications team repeated it every day with the aim of making the press conferences as boring as possible. After acknowledging the depth of the problem and promising a solution, there was no more news value in the daily ups and downs. This yielded tangible benefits, such as not having to face TV vans camped outside the office every day.

The engineers were aware of Vast Majority Sunday, but honestly did nothing in particular to prepare for it. As far as we were concerned, the goal *every* day was for the vast majority of users to complete their enrollments. For a while, we had no ability to guess in advance which days those would be.

The reality is that the days before and after Vast Majority Sunday were all about the same. But the media had been primed to expect the site to be working for the "vast majority" of users on that day. So when they tried opening the website on laptops in the studio and it worked a handful of times, they declared it a success. The crisis was over.

Section 2:

YOUR CRISIS TOOLKIT

2.1

Crisis Engineering Toolkit: An Overview

> The new firefighter, seeing black smoke rise from the ground and then at the top of the sky turn into flames, thinks that natural law has been reversed. The flames should come first and the smoke from them. The new firefighter doesn't know how his fire got way up there. He is frightened and should be.
>
> —**NORMAN MACLEAN,** *YOUNG MEN AND FIRE*

Our collective experience spans scores of interventions that successfully ended crises. We are usually operating in circumstances with little to no support, in a neutral or hostile organizational environment, with very little margin of trust or time. Therefore, our tools must be efficient and optimized for self-sufficiency. We have done our best to follow an empirical tradition, discarding what does not work and keeping the rest.

We've laid out this toolkit in practical detail, explaining how the indicators of crisis can be used to assemble them. We believe the reason these tools work is because they establish and encourage successful sensemaking, creating an environment that can accelerate, grow, and drive iterative actions that resolve problems.

This section is your tactical toolkit for managing and emerging from a crisis, and leveraging that crisis into enduring change. Ideally, you'll have had the chance to peruse this section *before* you're in crisis, but if you've frantically come to this page under more urgent circumstances, we've added a bulleted list at the end of each chapter for you to reference in a rush.

As crisis engineers who apply this toolkit for a living, we often shift roles. On one project Mikey may be focused on mapping the system, while next time Weaver focuses on mapping while Mikey prioritizes metrics. We use the second-person "you" liberally in this toolkit, recognizing that whoever you are, in any given crisis you will cover some roles while your teammates handle others. A crisis engineer needs to be familiar with the whole picture, even though they won't paint it alone.

YOUR MISSION: CONVERGE UPON ONE VERSION OF THE TRUTH

As we explained in chapter 1.1 (Sensemaking 101), **sensemaking** is the process of understanding the world and our place in it, especially in an unfamiliar situation that is ambiguous and maybe dangerous. We operate every day with inaccurate and incomplete pictures of how our organizations and their complex systems work—not so much because of incompetence, but because efficient System 1 thinking dominates. But when System 2 is needed, clinging to out-of-date models will prolong your crisis and lead to ineffective actions. Acknowledging that your mental model is wrong opens the door to fixing it, which can resolve the crisis and prevent its recurrence.

While sensemaking is a continual process, it's possible to apply it quickly. On our engagements, we usually enter with no familiarity with an organization, and achieve reasonable sensemaking by the end of day three or four. As your team or organization develops sensemaking maturity, it could, potentially, rapidly iterate and improve upon existing models in even less time. An investment in durable sensemaking during your current crisis is a leg up on your next one.

Often, sensemaking is understanding a process in a new and holistic way. For decades, teams may have operated serially and independently. It is very possible nobody in your entire organization knows how its entire process works. It's time to change that.

What's actually *wrong?* You need one accurate story—not for the sake of having a story to tell, but in order to be able to measure improvements against it. You can't resolve a crisis if you have no answer, or two different answers, to this question. It's okay to update this story as you gain information from taking actions.

YOUR CRISIS TOOLKIT

Books are inherently serial, but the next seven chapters describe steps a crisis engineer needs to take right away, *in parallel*, and continue to do throughout the crisis. These are:

Establish a Crisis Engineering Center ("War Room")

Sensemaking requires communication with many people who hold many perspectives. You have to talk directly to the people who know things. You have to make a safe space for them to tell you the known truths and discover new ones.

A very small team (no more than fifteen, but perhaps as small as two) needs to do this because the cost of information asymmetry here is potentially catastrophic. The team needs to be able to sit at the same table and converse to land at ground truth, run down changes in their information flow, and converge on actions.

In this chapter, we'll detail how to set up the workspace where you will construct consensus reality: the crisis engineering center.

Map the System

Sensemaking may include process maps, system diagrams, organizational charts, and/or data analysis and reporting. On our engagements, we usually create all these artifacts. These should coalesce into *one* shared picture/page that shows the path out of the crisis. Accuracy is less important than consensus—using a single, inaccurate map trumps using several more accurate maps.[1]

In this chapter, we'll walk you through mapping your system to establish a consensus reality.

Find Your People

The right people to resolve a crisis are rarely a preassembled team. In fact, it's generally our experience that the people with the most helpful information have never met before.

The right team will also shift over time. You need primary sources: people with their hands on the keyboard (whether they're typing code or policy). Whenever someone starts a game of telephone ("Olivia told me this morning that…"), you'll want to pull that source in directly. As your target moves, your primary sources will naturally shift, changing the makeup of the team.

In this chapter, we'll cover tactics for finding these people, as well as for leveraging other helpful people around the organization.

Take Novel Actions

If we haven't hammered it home yet…sensemaking is done by taking action, not by thinking or planning hard. In this chapter, we'll suggest some of the first novel actions you may want to pursue, including the surprisingly simple yet powerful step almost no one ever takes: Have you tried making a one-line code change?

Taking action doesn't mean taking every possible action

everywhere all at once. As your team, system maps, and data become more accurate, your action selection should become more strategic. Your ability to guess the impact of a new action will improve, and the size of the surprises they generate will decline over time.

In this chapter, we'll share stories of how to make these trade-offs in choosing actions.

Manage the Story

Once you've taken action and gathered data, you begin to assemble a shared, plausible story of what's happening—an essential step in sensemaking. This chapter shows how to bracket events by deliberately choosing beginnings and endings, using your observations and hypotheses to construct a coherent explanation that guides further action. The goal isn't perfect accuracy, but rather a plausible version of the truth the whole team can align around and improve upon. Your evolving story becomes a powerful lever for change by clarifying priorities, focusing attention, and ultimately driving better decisions.

Measure Progress

You've landed on a plausible consensus story of where you are. Now, how do you know where you're going? In this chapter, we'll detail how to measure your current state, model the impact of new actions, and then measure their actual impact. We'll also help you define health metrics that can warn of a future impending issue before things get too bad again.

Communicate in a Crisis

Sensemaking is social. It *is* communicating. In this chapter, we'll talk about the communications that you'll want to send out from the crisis engineering team.

You may have noticed that the world isn't ideal. You may do everything right, follow our advice to the letter, and find that it still doesn't work. At the end of this section, we touch on a hard question: *When should I give up?* Arguably, this should be the first chapter, but in our experience, nobody listens to our advice until they've first had some unsuccessful bites at the apple. Some lessons can only be learned the hard way. If you're feeling frustrated, head there.

Once you have established the seven foundations, and assuming you haven't skipped ahead to the part about giving up, the iterative and continuous process of sensemaking can proceed. The process will not be linear or orderly. Reframing a story will change the structure of the process map, which will cause the metrics to be updated, which will cause the team composition to change, which will cause the story to be reframed. But as long as you allow understanding to evolve and flow with new information, the distance between the system as it is and the shared mental model will decrease. Ideally, stress, chaos, and surprise will recede and leave behind a smoothly operating system that is adapted to the new environment.

2.2

Establish a Crisis Engineering Center

> Unless we are willing to escape into sentimentality or fantasy, often the best we can do with catastrophes, even our own, is to find out exactly what happened and restore some of the missing parts—hopefully, even the arch to the sky.
>
> —**NORMAN MACLEAN,** *YOUNG MEN AND FIRE*

The primary tool of crisis engineering is what we call a crisis engineering center. It shares a lineage with emergency response, incident command, mission control, and war rooms.[i]

We made up a new name ("crisis engineering center") because we suggest a set of practices that are expansive and complementary to formal incident command. We also want to highlight the shift in focus from mere coordination and communications to action-driven transformation of organizational behavior.

The center is the cornerstone of all the practices that come after

---

[i] Established best practices in these lineages can help if you are in the position of planning a crisis center for your organization. We suggest the U.S. FEMA's Incident Command System (ICS), and its forebear, California FIRESCOPE's ICS. FIRESCOPE was a retrospective sensemaking effort by wildfire responders following the disastrous 1970 fire season in California (sixteen deaths, $234 million in damages).

this section. It is the who, the where, and much of the how for generating sensemaking that leads to new actions, which in turn will modify organizational behavior and culture. It is the means to changing what a complex system does. It is the place where the circumstances of a crisis can be transformed into assets:

1. Fundamental surprise → discovering facts about reality
2. Failure of sensemaking → constructing a new consensus reality
3. Degradation, disruption, or complete change of core processes or outcomes → requirement for change, new actions
4. High visibility → being seen as blocking progress becomes uncomfortable, which reduces bureaucratic resistance
5. Rigid deadline or timeframe → speeds up the decision-action-consequence cycle, which leads to more reliable conclusions

COMPONENTS

The essential components of a crisis engineering center are:

1. A convening authority
2. A designated venue
3. Decision-making authority and access permissions
4. A single, well-known means of low-latency communication
5. An incident lead, and some number of experts and/or practitioners
6. A single shared journal of events and decisions
7. A broad, prominent announcement and kick-off ritual

A Convening Authority

A successful crisis engineering effort will, by nature and design, lead to: giving direction to staff, temporarily relocating personnel, temporarily reassigning duties, and temporarily expanding or contracting responsibilities. It will disrupt the normal working order of an organization. This disruption is a necessary driver of the transformation needed to exit the crisis. If continuing to operate in the same way could solve your current set of interlocking problems, it would have already done so.

A sufficiently powerful, influential, and/or highly placed **convening authority** must declare the crisis engineering center and assign staff to the effort. The authority must have the ability to influence operations and responsibility across all, or nearly all, of an organization. In the commercial space, it will usually be the CEO, a member of the C-suite, or a staff member reporting directly to a member of the C-suite. In government, it will be the head of the bureau's executive body, or sufficiently close to it, such as an agency deputy secretary.

If the organization contains sufficiently autonomous operating divisions, and the crisis is contained to one of them, then the authority should come from inside that division. If the crisis isn't contained to one division, then your divisions are not "sufficiently autonomous," and the authority must come from higher up.

All that matters is how the organization perceives the convening authority. Is it someone who can announce a new, overriding priority for everyone and make it stick?

People with this kind of convening authority are often reluctant to use it. They are probably responsible for wide-ranging trade-offs, so they will feel the costs of the disruption as well as the benefits. That's why they are an authority! This will serve as a litmus test of whether your problem is truly a crisis. The executive sponsor has to be ready to sacrifice other priorities, at least temporarily.

Recalling the five indicators of a useful crisis laid out earlier, most are the sort that can drive a convening authority into action.

Fundamental surprise, failure of sensemaking, disruption of core processes or outcomes, high visibility, and rigid timing constraints are all things that, in our experience, prompt a powerful leader to declare a crisis. It is really this simple, but it may not be easy.

Regular order is the standard process by which work moves through an organization—governed by established roles, approvals, and timelines. The convening authority is the primary means to signal **nonregular order** to the organization at large, allow the reassignment of staff and responsibilities to the crisis engineering center, and declare a new global top priority for the organization.

It is pointless to try to gin up a crisis engineering process without an engaged authority figure. To do so amounts to running an insurgency. That calls for a whole different book full of different tactics. When crisis engineering is possible, authority figures will already be motivated to act. You might then propose a crisis engineering center, and perhaps even tee up a draft announcement. For better or worse, doing this may make *you* the incident lead, which we'll discuss shortly.

A Designated Venue

We call it a crisis engineering *center*, in part, because it must have a physical location. The work has to happen somewhere, and it must be concentrated. The members of the crisis engineering team must be compressed together to emphasize the social aspect of sensemaking. They also need access to the broadest set of facts available, including the wide array of signals we pick up from one another's body language and other behavior. Critical information will be lost if individuals are scattered around. Drawing members of the crisis engineering team into a shared sense of purpose, a shared understanding of the story of what happened and what we are doing next, and a shared stream of available facts is critical if we want to accelerate sensemaking and increase the tempo of actions taken to affect a complex system.

So, get a room. The convening authority will be able to assign spaces like a conference room or section of a cube farm. The room needs places for people to sit, power for laptops, and chargers. A whiteboard is good, and it doesn't hurt if there is a big screen or two on the wall. It will need network access, and Wi-Fi with the network password tacked to the wall. Videoconferencing equipment is nice, but not required. Anyone in the organization should be able to find and enter this room (e.g., it is not behind a restricted keycard).

Unusual or exclusive spaces can be good, as long as access is easy. If you have to use a space that is part of the regular workday, like a common area or standard conference room, then you need to make it obvious at a glance that something new is going on. Taping a sign to the door is a start. People will not react to a crisis effort if it looks the same as every Monday team meeting.

If the organization is currently experiencing "remote work," it's best if the convening authority approves and mandates travel to the center for any staff identified as necessary by the crisis engineering team. If your organization is fully remote and has no physical locations, create one at a coworking space or a hotel near a major airport.[ii] All things are possible if the crisis is urgent enough.

If all else fails, a room with videoconferencing equipment that's permanently dialed in to a dedicated virtual meeting is a *little* better than nothing. But not much. Time zone complications and the loss of communications fidelity will impede sensemaking. An important part of the group consensus you are building is salience and urgency: "We all know this is an emergency." Your efforts will be undercut every time somebody dials in late while in line at a Starbucks with their camera turned off. All in all, it makes everything harder. You are trying to run a sprint at the very limits of your capabilities, and you're starting by tying a sandbag to each ankle.

We can't reiterate enough how important it is to get people into one place. The more unusual this is for your organization, the more you will benefit from the signal that something new and serious is going on.

---

ii For months, the lobby of the Columbia DoubleTree hotel was a de facto extension of the HealthCare.gov exchange operations center.

Decision-Making Authority and Access Permissions

The speed of sensemaking and problem-solving is driven by the speed that you take action. Every single thing that postpones or prevents a crisis engineering center from deciding on an action and taking it impedes progress.

One of the most common impediments to action in complex systems is a lack of authority. Practitioners are, in normal conditions, concerned with the allocation of authority, risk, and blame. Actions are postponed pending decisions or authorization by other bodies in the organization.[iii] These delays must be eliminated or minimized in a crisis engineering effort.

It's best to maximize the number of decisions that can be adjudicated from inside the crisis engineering center.

One solution is to bring a sufficiently powerful overseer into the room who understands that her purpose is to grant permission and make decisions on the spot. She will be accepting a lot of risk. This usually requires someone who has nothing more to lose. We have seen it done by senior managers who were about to retire, and by political appointees who know their fate is sealed anyway. We've seen the U.S. Department of Defense use a widely respected general near retirement for exactly this purpose. CEOs will do it when they realize that failure will end the company.

Other times, you will have a cloud of managers who all calculate that they can survive the crisis if they don't touch it. In that case, the best bet is for them to hire outside experts and delegate operating authority. The fancier and more expensive the consultants the

---

iii We love Anthony Downs's explanation: "This consists of rigidly applying the rules of procedure promulgated by higher authorities. Instead of 'playing it by ear' and adapting the rules to fit particular situations, many conservers eschew even the slightest deviation from written procedures unless they obtain approval from higher authority. Thus, rigid rule-following acts as a shield protecting them from being blamed for mistakes by their superiors, and even from having to obey any orders that conflict with 'the book.' This attitude of rigidity, plus the delays involved in obtaining official rulings for unusual situations, create the conditions that have become stereotyped as 'the bureaucratic mentality' and 'red tape.'" (See Downs, *Inside Bureaucracy*, 100.)

better, because if things go wrong the managers want to say they did all they possibly could.

Exactly how you accumulate the necessary autonomy will be slightly different every time. What is important is that the crisis engineering center can rapidly (in a few minutes) make any decision, or get any approval necessary, in order to proceed with the next action.

A Single, Well-Known Means of Low-Latency Communication

We cover communications in a later chapter, but inside the organization, anyone must be able to easily reach the center, to ensure new information can reach the crisis engineering team.

This communication channel must be accessible, by anyone, regardless of their status in the organization, and ideally without requiring specific tools or authentication schemes. It must be low-latency, meaning anything communicated to the center reaches everyone in it *without delay*. The timeliness of information is paramount.

This rules out email. We don't recommend a group text chat or Slack, either, unless someone in the center is tasked with monitoring it and reading messages aloud as they arrive.

The channel must also have primacy, meaning there is *one*, highly publicized, monitored way to tap into the center with new facts. Having multiple ways to reach the center will breed confusion and introduce lossiness to the process of acquiring information. This will be lethal to the sensemaking going on in the center.

In our experience, this is a case where the simplest thing that could possibly work is best. An old-fashioned, always-on (24/7) phone conference in the crisis center room is great. A speakerphone is low-latency and commands attention. (It's hard to ignore something squawking in the center of a room.) This is good. A phone bridge lacks video, which is also good, because it doesn't provide either party the illusion that it's a substitute for being in the center. There is a natural tendency to ask someone to come in person when communication becomes too complex for a phone call.

An Incident Lead, and Some Number of Experts and/or Practitioners

The convening authority must appoint the incident lead upfront; the other team members will be in a state of flux as the effort progresses. We go into more detail about finding the crisis engineering team members soon, in chapter 2.4 (Find Your People).

The incident lead is the tactical leadership role over the crisis engineering center. The exact title is not important; we've seen "incident lead," "pit boss," or even "response chair."

The lead doesn't have to be from management and doesn't have to be a person already recognized as powerful by the organization. They *do* need to be seen as someone who cares about solving the problem more than anything else. People will contribute their best if they think it is helping solve a problem. They will not contribute their best if they think they are helping some annoying person get promoted.

The lead must understand they are responsible for keeping the sensemaking loop running. Some people know how to do this by intuition. Of course, it will help if they've read this book!

We've seen many successful leads who have worked in theater (the show must go on!), as first responders, or in the military. What these jobs have in common is that it's obvious that doing *nothing* is not an option.

For crisis engineering centers that will operate around the clock, you'll need at least two leads who trade off in shifts.

As for the rest of the team, the key step to take now is to start a list of potential team members. It may be frighteningly empty at first, but you'll soon need a source of truth as to who is (and is not) officially invited.

This list needs to have names, information on how to contact them, and a suggestion of their role or areas of expertise. Keep it centrally accessible, like on a whiteboard or the front of the journal we explain in the next section.

We try to be as definite as this example:

SAMPLE ROLES & CONTACTS

ROLE	ASSIGNED	EMAIL	CELL PHONE	SHIFT
Incident Lead	Duncan Idaho	d_idaho@caladan.gov	555-555-5555	Daily 0800 - 2000
Incident Lead (alternate)	Thufir Hawat	t_hawat@caladan.gov	555-555-5555	Daily 2000 - 0800
Communications Lead	Gurney Halleck	g_halleck@caladan.gov	555-555-5555	
POC: Planetology	Dr. Kynes	dkynes@empire.gov	555-555-5555	
POC: Bene Gesserit lore	Lady Jessica	j_nerus@caladan.gov	555-555-5555	
POC: Back-stabbing	Dr. Yueh	w_yueh@caladan.gov	555-555-5555	
POC: Muad'Dib history	Princess Irulan	icorrino@empire.gov	555-555-5555	

A Single Shared Journal of Events and Decisions

The center must have a single, shared journal that is updated with events and decisions as they take place. Remember that sensemaking is *retrospective*: It happens as we look backward in time and come to an understanding of *what has happened*. If there is no easily accessed record of what things have happened, what decisions the team has made, and what actions the team has taken, then sensemaking will not occur. Confusion and chaos will reign. Rather than making progress, you will find yourself repeating circular arguments.

The incident lead is responsible for ensuring the center discusses what the team knows, and decides what action to take next, at least once a day. The incident lead must also ensure that the discussion and decision are written down.

When a hypothesis is made or an action is tried, write it down, with the time stamp. What happened? Write that down. Continue the log for the duration of the crisis response. You might also keep a "sandbox," or a "backlog" list of ideas to try next and/or to come back to after the crisis.

A large whiteboard will work for the journal (if you're confident no one will erase it!); a single shared collaborative document works as well. There must be only one, and it must be accessible to every person on the team at all times. Turn on the document's edit history to discourage historical revisionism. Crisis centers that email copies of "todays_journal_final_final(2).docx" will perish. So will the ones who fragment the "journal" into a thousand Jira tickets organized in no particular order.

The incident lead is responsible for the accuracy of the journal. Every time a notable event occurs in the complex system, every time the crisis engineering team meets, and every time the team takes an action, the lead should update the journal in reverse chronological order (newest items at the top).

Inevitably, the team will disagree on what happened in the past. With a journal of events, actions, decisions, and information, the team can reach consensus about *what happened* and *what we did*. Without a record, they cannot, and consensus reality is impossible.

A Broad, Prominent Announcement and Kick-Off Ritual

The announcement of the crisis engineering center is a particularly important piece of organizational magic. It must come from a sufficiently powerful leader, as explained earlier, if you expect anyone to heed it. It should be conveyed to the entire organization.

The announcement must do at least two things. First, it must declare that there is *a thing*—a new crisis engineering effort. Give it a name, and make it clear that it is now The Most Important Thing. The name of the effort doesn't matter, as long as it's recognizable. Second, it must convene staff into that new thing, by *mandate*. If the team needs you, you drop everything else and come.

Declaring to the entire organization that the center is temporarily the top priority signals to everyone that something new and different is happening, and it is important. The disruptive nature of a crisis engineering center is not an unfortunate side effect—it is a big part of the center's purpose.

This announcement can take more or less any form appropriate to the organization, as long as it accomplishes the above. In some organizations, there may be formal means for declarations like these. Use those. It's also fine for the announcement to be a company-wide email from the convening authority, similar to a CEO's end-of-year missive.

The announcement should contain *all* of the following:

1. Declaration that there is a crisis engineering effort that temporarily overrides all other priorities
2. The name of the center/effort
3. The location of the crisis engineering center
4. The primary problem(s) the effort seeks to solve
5. The name of the incident lead
6. Any authorities delegated to, or adjudicating decisions for, the effort
7. Specific contact details for reaching the crisis engineering center

If you know (say, from a past crisis) how you'll keep the organization updated (e.g., on a status page), share that now. If you don't, share this later.

THE KICK-OFF RITUAL

The announcement of a crisis engineering effort by the convening authority should be accompanied, if at all possible, by some sort of kick-off ritual. Starting the crisis engineering center is part of how you start framing the story of this crisis. (We cover this storytelling strategy, which we call bracketing, in chapter 2.6.)

When a crisis comes along, everyone in the organization is affected. The initial shock destabilizes people's sense of identity. This is what makes it possible to break or create habits, which is our whole theory of why change is possible in this window. Almost everyone's natural instinct is to get involved and contribute in some way. If you supply a plausible story for what caused the crisis and what's likely to fix it, along with some encouragement, you can get the best out of people for a time.

Most executives understand this much on instinct. We have all been to some version of the kick-off meeting when management wants to start some new initiative or make a leadership change. It was probably in an unusual venue at an unusual time. A hastily planned gathering in a makeshift venue can actually be more effective at communicating there is a problem. The de facto kickoff meeting for the HealthCare.gov rescue effort took place late at night, in the lobby of an empty office building in Virginia. Hundreds of staff had been summoned to Virginia on short notice, with no room ready to hold them all.

It's important for a kick-off ritual to include a call to action that includes expectations for everyone in the organization. The foundation of all sensemaking is each person's sense of their identity and role. For the people who are going to be part of your crisis team, these things need to temporarily change. For the people who do not need to contribute to the crisis team, they may need reassurance that their role and identity are staying the same.

Many managers fail to do this. The result is eye-rolling as everyone shuffles back to their desk with a sense of "yet another meaningless announcement." People are going to process the event with their friends, at lunch or in the hallway, and a constructive result is critical to your future success. You need to impress upon as many people as possible that:

- There is a new threat.
- The organization is doing something new to address it.
- In time, everyone will be part of that "something new."

The call to action does not need to be precisely specified yet. Yes, you will have people who immediately raise their hands with a question or send you an email asking about minor details of a plan that doesn't exist yet. It is not necessary to answer all these questions. Contrary to the widespread assumptions, it does not harm leadership's credibility to communicate uncertainty.[1] It is worse to confidently declare a plan and be forced to backtrack later. Worst of all is to pretend to have a plan you aren't sharing. No one will be fooled, and nothing more that you say will be trusted.

KICKING PEOPLE OUT OF THE CENTER

The crisis engineering center should have its identified team members, invited special guests (like a designated decision-maker), experts dropping in by invitation, and the occasional appearance of an unexpected guest with information to share. Extra bodies are distracting and unhelpful. They increase the complexity of communication, as each additional human needs time to speak, to connect to other team members, and to be caught up when they step out. At a certain, and early, point, a group is too large to have an effective discussion. We have advice for getting unnecessary audience members to go away.

The most successful strategy is to bore them with technical detail until they leave on their own. The incident lead should ruthlessly focus on the weeds of the potential problems and solutions, with the technical experts in those areas. This should be happening regardless of the presence of unhelpful outsiders, but the level of technical detail can be amplified with this end in mind.

In a pinch, invite audience members to present on how the internet works or lecture on the history of a specific relevant legal precedent. A bold and motivated incident lead might use the Socratic method to publicly ask the unwanted audience member a highly technical question, causing them to suddenly remember an urgent meeting across town.

If they still won't leave, and they are sufficiently senior (or squeaky), you may have to sacrifice someone to be their very special crisis envoy. This person will be responsible for briefings and updates at whatever cadence and depth the leader requires to be able to get those updates from the comfort of their own office.

Of course, the most effective way to get rid of a peanut gallery is to effectively manage and end the crisis at hand, so *everybody* gets to go home.

If your situation room is mostly filled with spectators and individuals with explicit incentives that are not aligned with ending the crisis (such as increasing their personal visibility, or launching their pet project under the guise of the crisis response), *the crisis is over*. Now it's *your* turn to leave the room.

Now that the crisis engineering center infrastructure is in place, it's time to start the crisis engineering effort. The following chapters need to be done *in parallel*; the first of these parallel steps will be to map how the world actually works.

CRISIS CHEAT SHEET

A **crisis engineering center** is a physical place where team members convene to develop and update the shared consensus reality.

To create a crisis engineering center, you need:
- **A convening authority**—A senior official with plausible authority needs to declare a crisis, point to a place as the crisis engineering center, and convene staff for the effort.
- **A designated venue**—An actual, physical room where people get together in person. In a remote or distributed workforce, this means people traveling to one location. We're quite serious about this: Remote crisis engineering efforts are significantly less effective. Ideally, this is a centrally

accessible location where anyone with information can drop in.
- **Decision-making authority and access permissions**—As many decisions as possible need to happen in the crisis engineering room, without external approval chains, in mere minutes. This may include purchases of up to some threshold, the ability to change firewall rules, and the ability to pull in (or kick out) arbitrary staff members (including contractors) as needed. For those decisions that cannot be delegated to the crisis engineering team, the decision-maker needs to be on standby to adjudicate decisions immediately. The team also needs broad access to buildings, humans, and machines.
- **A single, well-known means of low-latency communication**—We strongly prefer a phone bridge, but a closely monitored instant messaging channel can also work. It should be well publicized with a low barrier to entry, so people with information to report, such as a support desk manager sharing an unusual volume of new tickets or a remote engineer reporting a related bug, can join easily. This channel should be open and monitored 24/7.
- **An incident lead, and some number of experts and/or practitioners**—Team members will change over time as the needs of the crisis engineering effort evolve; keep a centrally accessible list of current team member names, contact details, and rough roles and responsibilities so everyone remains on the same page.
- **A single shared journal of events and decisions**—This can start as a whiteboard or piece of paper on the wall of the crisis engineering venue, though it ultimately should be in a more resilient location like a shared online document with edit history. Record each decision and action taken, with time stamps.

- **A broad, prominent announcement and kick-off ritual—** This announcement, from the leader with convening authority, needs to declare a crisis, name the team lead, identify the location of the crisis engineering center, and detail how to reach the communication channel. If there is a place where status updates will be provided more broadly (particularly important if you have a large, impacted user base that will otherwise overwhelm your support lines), announce that, too. This communication needs to make clear that the crisis engineering effort is the number one priority: Managers must allow their reports to respond to any requests.

2.3

Map the System

> The parachutes were made of nylon because grasshoppers like the taste of silk. In a modern tragedy you have to watch out for little details rather than big flaws.
>
> —NORMAN MACLEAN, *YOUNG MEN AND FIRE*

After you've established your crisis engineering center, one of the first things to do is map out your systems. The goal is to cocreate a plausible shared reality of how information *actually* flows across the organization, and your current best guess as to how changing one part of the process influences other parts downstream. This shared reality will encompass both humans and machines.

We need to tell you something you're not going to like, but is actually your most powerful tool for transformative change: The map that you have now of the systems around you, whether in your mind or on paper, is *wrong*. This is especially true in a crisis (or it wouldn't have snuck up on you), but don't feel too bad—it's also true the rest of the time.

Steps you would bet your life are happening reliably are not happening at all. Policies are being implemented in ways counter to their intent. Somewhere, the right hand loses what the left hand gives it. A process that takes four weeks today could actually be done in seconds, because nobody in the last thirty years ever thought about it end to end.

How does knowing this help you? Because if you create a more accurate map *now*, with the benefit of the unprecedented access and rapid feedback provided by the crisis, you have the proverbial keys to the castle for changing the organization. While others continue operating in the dark, or on incorrect information, you'll have the closest thing to reality, which you can use to rapidly take actions that increasingly work as you expect them to.

People can (and do) sit in an isolated conference room and argue, forever, about what's really happening outside, without ever actually checking their assumptions. This argument is much harder to continue when everyone is forced to look directly at the same underlying code, algorithm, or control panel. People *hate* cognitive dissonance, which is when your beliefs do not align with your perceived reality; they want their consistent view of the world back. Therefore, the social process of mapmaking is also a wildly effective tool for quickly and permanently *changing minds*—a tool you can use to your advantage.

A plausible map won't only help you resolve the immediate crisis indicators; it can also help you orchestrate outsized organizational change within the crisis period, and significant (if diminishing) changes thereafter.

THE GOAL IS TO CONVERGE ON ONE SHARED VERSION OF THE TRUTH

Mapping can help create a consensus reality and give you a shared understanding of how the system operates, where its weaknesses lie, and what success looks like. You'll then direct actions against this reality to change it for the better. When your actions make things worse or have no effect, you'll update the map to reflect what you've learned about how things *actually* work.

At a high level, this consensus reality needs to span the following:

- **Process**—How things actually work today, including requirements and priorities

- **People**—Who is involved, who has authority, and who carries out specific actions
- **Data**—How the business process is represented in systems and/or code, and how data flows change in concert with real-world actions

This shared understanding sets the stage for the following:

- **Objective Measurements**—How to track progress toward ending the crisis, and identify what worked (and what didn't) against a shared understanding
- **Action Plans**—How the team will align on what actions to try next
- **Reimagining the Possible**—An updated map can practically roll out the red carpet for a change or an initiative that seemed impossible mere minutes before

This map needs to be *plausible* to the crisis engineering team members and those familiar with the parts of the system it reflects. It does not need to be exact or perfect.

Begin by tracing a single end-to-end flow, for example:

- A transaction, with a single, authoritative count of customers or lost orders, and/or a single, authoritative definition of "downtime" or "fraud"
- A customer interaction, with a single, authoritative definition of what counts as an interaction and whether it was positive
- Manufacturing a product, with a single, authoritative number of defects, clients impacted, and/or cost
- Processing a claim, with a single, authoritative definition of "claim" and "backlogged claim," and count of backlogged claims
- Treating a patient, with a single, authoritative number of adverse outcomes, patients impacted, and/or cost

- A natural or technical disaster, with a single, authoritative definition of "recovery" and means to measure proximity to it (e.g., "electricity and water restored to N homes")

There's no reason you can't map your systems just as well outside of a crisis (though it can be slower going to find the truth), but no one ever does. Maps made during normal order will be based on opinions (things people have said) rather than facts (observed with the mapmakers' own eyes).

HOW BIG OR SMALL IS THIS MAP?

A great and terrible thing about sensemaking is that the process has no beginning or end. A great and terrible thing about systems thinking is that there is no obvious boundary that defines the "system." If you are stuck on a problem that seems impossible to solve, it may be the case that your current model of the system is missing something. It could also be that you're including something you shouldn't, and the problem ceases to exist when you cut it out. Let's look at clues that indicate you should change the scope of your analysis.

A Simple Model of Control

Cyberneticists and systems-safety people like to work with simple abstractions. One of these is the "control loop." The idea is that all the intentional actions happening anywhere in a complex system can be attributed to an agent that operates on some internal rules (a **controller**), its inputs (**sensors**), and its available actions (**actuators**). Controllers can be machines running rigid algorithms, people running rigid algorithms, or people exercising discretion.

In principle, any system can be mapped as a set of interconnected control loops. Any piece of the system that matters must be performing the function of a controller, a sensor, or an actuator. Moreover, it must be part of a loop that contains all three. This can help you find pieces you have overlooked.

When to Zoom In

Picture what you usually think of as your system. This is probably a diagram you habitually draw on the whiteboard.

Where are there components that take in a set of information, make a decision, and take one of a set of actions? Each of these is a control loop of its own. Loops that are made of machines are easy to find, because their communication patterns are easy to find. A computer program cannot pull another computer program aside in the hall and have a quiet word about the disability application it just sent over. It can only send bytes into the channels you gave it.

Loops that are machine-human hybrids take more work to find, because there is no mechanism for automatic discovery. You have to do the tedious, time-consuming work of watching people do their jobs, or at least asking them how they do their jobs (and interpreting their answers, the art of which is a lot of this book). Loops that are entirely human are uncommon, because the actual production data being acted upon is almost certainly stored in a machine. But it's not impossible! Marina worked on a process at the Department of Veterans Affairs where tasks and ownership were defined by moving empty file folders around the building. The papers that used to be in the file folders were in a computer database, but it was the location of the folder that indicated the current status. This modernized process had to be understood as entirely human, since the computers were serving no function except to make the folders lighter.

Each control loop that is thus discovered can be examined as its own autonomous system to some degree. We can then evaluate it for all the known principles of good systems: Does its controller have well-defined goals? Does it have sufficient sense data to pursue those goals? Does it have blind spots? Are there times when a corrective action is knowable but cannot be put into action? Does the controller have enough resources to handle the set of decisions it might need to make?

You will know it is time to zoom in when you find yourself diagnosing problems that you can do nothing about. For instance, in government systems there is frequently no executive who can make a meaningful change to budget or head count, which may be set by law. There is no point admiring this kind of problem from a crisis engineering perspective.

When to Zoom Out

The systems and cybernetics approaches scale both ways. Imagine any complex system you know, like the postal service or your favorite restaurant. You have a sense of what things are under its control and what things are not. Now ask yourself, "Who and what will intervene to fix it if things go completely off the rails?" You will discover that what you perceive as "the system" is actually embedded in a much larger system. The way to learn about the larger system is to find answers to what can be done to the *smaller* system—critically, *without its cooperation*—and by whom.

Let's say you are responsible for a small product at a large tech company. You have an idea of whether things are going well or poorly at any given time, which derives from the factors that are important to you, like your performance bonus. The company is also collecting data on things that matter to it, such as "monthly active users." If this number becomes very small or very big, the VP of Product will make a decision and cause actions outside your control. That is to say: You are the subject of a higher-level control loop.

Keep on looking—there's no requirement that your system be controlled by exactly *one* other system. Besides your VP of Product, there are other power centers in your company looking at other things. For example, human resources may be making observations that have nothing to do with the size of your user base. They intervene according to their *own* set of conditions, which are also outside your control. You probably also have an analogous department for seeing to the needs of the machines, called something like "data center operations" or "IT" or "SRE."

You are the manager of this small product, but the VP of Product, human resources, and IT each have the ability to do things to your system without your cooperation. Each of them is a controller. If their actions can create a problem you care about, you need to add them to your mental model of your system.

The important thing to remember is to add not only the *controller*, but the whole *control loop*: their internal rules, their inputs, and their potential actions. As the one being controlled, you already know something about their potential actions. You may have no information about their inputs or internal rules; you will need to learn about them.

Widening the scope of your systems thinking should be done carefully, because adding complexity that isn't necessary or relevant will only make things harder. *You will know it's necessary to zoom out when a poorly understood external controller is causing outcomes you don't like.* In this situation, you must understand what information the controller receives and how it makes decisions.

Why Systems Tend to Self-Encapsulate

Everyone wants freedom to act, and no one wants to be acted upon. This causes human controllers to make themselves difficult to predict. In our tech-manager example, it's unlikely that there is a simple algorithm that determines what number of monthly active users is too low, and what happens when you get there. In the places we have worked, there is an amorphous committee of managers and executives that functions as the "product strategy" controller, and it has no clearly defined goals.

The actions available to that committee are guessable. They may merge your product into another one—generally not a good thing from the perspective of the smaller entity. They may shut you down altogether. Or they may decide to invest more attention, money, or talent. Even that is usually unwelcome, since it will come with higher scrutiny and less autonomy for you. Thus, you are motivated to prevent the higher control system from ever getting involved in the first place.

This simple dynamic explains an awful lot of the behavior of bureaucracies and systems. Managers tend to want to maximize their autonomy, which means minimizing the interference from *their* manager. There isn't much they can do to influence the goals or thought processes upstream. Nor can they take away much of their supervisor's freedom of action. That leaves only the sensory data to be manipulated. Less information sent up the hierarchy means more predictable actions coming back down. That is why, as time goes to infinity, all executive dashboards converge to "Everything is OK, all the time."[i]

A cybernetic interpretation might be something like this: Any given manager wants to reduce variety in its upstream controllers, and they can do so by reducing the variety in their input data. Anthony Downs describes it as the **Law of Counter-Control**: "The greater the effort made by a…top-level official to control the behavior of subordinate officials, the greater the efforts made by those subordinates to evade or counteract such control."[1]

The tendency of systems to self-encapsulate helps organizations scale up. Understanding this tendency can help you better interrupt and/or leverage it to your benefit. It also adds a lot of internal overhead. A manager with a large portfolio will think up reporting requirements for her submanagers. The reports will be found inadequate. The reports will be supplemented with an in-person review and cross-examination. The submanagers will spend weeks preparing for the in-person review. And so on.

---

[i] At the instant this sentence was written, the U.S. government's IT Portfolio Dashboard at https://itdashboard.gov reports that 90.23 percent of federal IT projects are on schedule and 91 percent of performance targets are being met. This is risible to anyone who has any contact whatsoever with federal IT.

> It took less than a year after the founding of USDS before Mikey found that the regular "deep dives" with the project teams contained little to no actual information. Mikey's goal to optimize the allocation of scarce people across projects contradicted the project manager's goal to protect their team from changes. Therefore, those teams quickly began to manipulate the sensory data they provided to Mikey. When incentives are in direct conflict like this, it's not realistic to expect open negotiation to work.

MAPPING INFORMATION FLOWS

Perhaps you have a clean, printed system diagram full of neat boxes and arrows hanging on a wall. Great! Flip it over—now you have some blank paper to draw your actual process on. Real maps are messy, annotated, and ever-evolving. It is our experience that pristine maps, on the other hand, never accurately reflect reality.

This approach applies equally to technical systems and the business processes they support—whether you're processing claims, fulfilling orders, resolving customer tickets, or troubleshooting a network outage.

On Layer Aleph projects, two of us physically travel together whenever possible to build the initial map. We then constantly iterate on it as a team as we learn new information and as the people with whom we share it point out errors, omissions, inconsistencies, and exceptions. The map itself may reference external objects, like screenshots of a user interface, an example (but real) data import file, or a database schema.

The incident lead is the ultimate owner of the map, though they may delegate the actual updating of it to someone else. If delegated, the assignment is not to go off alone and come back with a report a week later; it's to capture and uncover reality in a highly social way. Extroverts are well suited to this task.

Store the map in the crisis engineering center where the whole team can see and modify it. We genuinely suggest using paper at first, and worry about diagramming software, or other formalization, later.

Don't Agonize over the Starting Point

If you have a perspective on how zoomed in (or out) you'd like to be, great. If not, just take a guess, and see where it leads. It matters much more that you start than exactly where you start.

Mapping the realities of your system starts with basic but often overlooked questions:

- How does the thing work?
- No, really—how?
- Where does a user or customer enter the process?
- What does this look like to an employee?
- What do they encounter along the way?
- What are they trying to accomplish?
- What does success look like?
- What can go wrong?

These questions apply at multiple system perspectives:

1. **User Experience**—What do users see, hear, or do? What the outside world sees is the actual truth, not what people *inside* the organization perceive to be the users' experience.
2. **Business Process**—What workflows or steps must occur? What exceptions are there? What flexibilities are allowed by law/policy, and which are allowed by the system? What are the business priorities and requirements?

3. **Organizational Context**—What teams, policies, or incentives shape these processes? Who can make what decisions? Who has authority over decisions, by subsystem? Who influences those decisions, formally and informally? What are their risk and incentive frameworks? Who can take direct actions on the machine parts of the system?
4. **Technical Systems**—What software, hardware, and integrations support these functions?

   For technical systems, make sure to cover:

   - What kind of computers? How many? Where?
   - What platform and/or languages?
   - Where is the dashboard and/or monitoring?
   - How does it connect to the other system(s)?

   Questions that start to reveal the truth include:

   - Who can change it? Who can authorize changes?
   - What does the policy or training manual say? What does the code that implements that policy or practice say?
   - How often does it change?
   - How do you know if it is broken?
   - Why does it break? What do you do to recover?
   - What's the next step? What happens before this?

As you discover so-called bugs or errors or issues, it's important to ask operators whether they're actually important. Many alerts are meaningless, while others are mission-critical; you may not be able to parse the difference as an outsider, but the people dealing with them every day can easily tell you which ones they click past mindlessly and which cause them to urgently pick up the phone.

A particularly effective strategy when mapping a system is to pay attention to the space between the silos: steps where different teams or systems hand off. This space tends to be the least known and least resourced. Data transformations, deletions, batch limits, duplications, errors, and more may be found at every handoff, and may explain a *lot* of what's going on—but because no one is in charge of the space, no one is aware of the issue or complexity. It's also often one of the easiest places to make changes that improve other parts of your flow, because there is no one to defend the status quo or argue against change.

The opportunities between the silos are why it's critical to follow bugs, network traffic, claims, orders, or whatever else you're following from start to finish through *real* flows, and not to accept narrated, abstract reports or explanations found in user manuals.

Building a more accurate map also means uncovering messy, human realities:

- See things with your own eyes. Visit the factory floor, observe customer interactions, and walk through physical spaces. Physically go to the call center, if there is one.[ii] Visit anything called an operations center.
- Talk to real users and frontline employees. They often hold pieces of the puzzle that leadership, management, and documentation miss.
  - If someone is resistant to sharing, see if you can nail down their reason. Do they need permission from their boss? Can you promise anonymity? Is there someone else around who'd be more eager to share? You're up against the clock, so arguing or forcing someone to contribute to your map won't work. Reticence is its own form of information. If multiple people won't talk to you, there's something important you don't know.

[ii] We have visited call centers that had no idea they were call centers (only that they were cube farms of employees who happened to have phones), and empty call centers that everyone else thought were open.

- Ask people to tell you a mystery. Encourage the people you're shadowing to share surprising or confusing process details. That question they don't understand or the mystery about why a status is captured differently in two different systems are often clues to process mismatches or solutions.
- Mix up who you ask. If there are a hundred help desk agents, don't send all your questions to one of them. There is a lot of value in asking two people the same question and seeing if their answers line up.
- Investigate your inputs. Is your call center operational? Are you capturing all relevant data? What inputs might you be missing entirely?[iii]

Just as a clean diagram is not an accurate depiction of any real-world process, all answers are lies unless you see them with your own eyes in a live system.

Use the map as a living, breathing tool in the crisis engineering center. Refer to it during discussions, update it as you discover new realities, and improve it as you take action and discover that all did not go exactly according to predictions.

---

iii Physical mail, informal or undocumented email chains, spreadsheets, and handwritten notes are common missing inputs.

SHARPEN OBSERVATIONS WITH DISTORTION-RESISTANT QUESTIONS

There is a critical difference between machine controllers and human controllers. Machines do not fight to preserve or expand their autonomy. Humans do. The only difficult part about finding out how a machine controller makes decisions is getting access to its source code. This is why the crisis engineering team needs broad system access, as granted by the convening authority. (It's also why the team needs to include people who can understand source code.)

Human control structures are often opaque on purpose. Executives and committees hate to give visibility into how they make their decisions, lest people figure out how to manipulate them—which is, yes, what you are ultimately trying to do with a map. It's best to figure them out by observing their inputs and actions as much as you can. Even when they are not actively trying to hide the ball, it does not work well to ask people directly why they do what they do. This will get you a System 2 answer, which is a rationalization, as opposed to reality. Unlike you, they are not aware that most of their decisions are made by their sensemaking System 1 on autopilot.

As much as possible, you need to make your data collections resistant to distortion.[2] This means asking questions that are specific, are quantifiable, and minimize judgment. This is a skill that will serve you well in and out of a crisis.

Instead of asking "Are there cost overruns on any of your ten biggest IT contracts?," ask for the total dollars that have been obligated as of the end of the last fiscal year and for copies of the original RFPs. Obviously, this is more work for you. But a person who might have been tempted to lie is less tempted when their answers can be fact-checked. Moreover, people who mean to answer honestly are deprived of shortcuts that might cause the answers to be lies anyway. When asked whether cost overruns are occurring, they might reason that an overrun requires an exception report to be filed, and no reports were filed, therefore no overruns happened. This is a bulletproof defense to a bureaucrat, and does not get you the right answer.

People don't want to ask dumb questions or say they don't know. They especially won't do this in front of superiors, strangers, or direct reports, and especially not in a group meeting. One-on-ones can help make them more comfortable, although if you outrank them, or they think you have more influence than they do, the power differential might be too much for them to speak openly. This is fundamental and must be treated like it is as inescapable as, say, gravity.

As you map, maintain a stance of curiosity rather than critique. Remain calm. Resist the temptation to assign blame or impose solutions prematurely. Your role is to listen, observe, and document without judgment. If the opportunity arises to praise someone who has made a mistake and admitted to it, trumpet it from the rooftops so others will hear. This approach fosters trust and ensures the map reflects reality rather than aspiration.

If you truly want only to understand a problem without inducing any changes, then you can only monitor the machines. The unemployment work queues that we describe later were a good point of leverage, because they can be measured using SQL queries and a postage scale.

When all else fails, use proxies. We have come across many log files that we were forbidden to read because they contained sensitive health information or some such. But no one can use privacy rules to prevent us from checking the size of the file. The rate of growth in an error log tells us something about the health of the system, even without looking at its contents.

MAPPING AS A TOOL FOR TRANSFORMATIVE CHANGE

When going through this process, you will discover uncomfortable truths. In particular, you will expose the fact that the formal command hierarchy does not have as much control as it thinks. Running code supersedes whatever is written in local policy, and local policy trumps regulations and laws. Habits of field operators override the employee handbook. Understanding these disconnects is key if you are to ever predict the results of taking an action.

Take this example from TD Bank:

> *TD Bank faced $3.1 billion in fines for failing to prevent money laundering. Banks are required to report the name of the individual making a transaction against the name of the account holder. Because their system automatically filled in the transactor's name with the account holder's name, this protective measure was missed hundreds of times when criminals were laundering funds.*[3]

Compliance had a policy. The computer screen implemented something different. If compliance had seen this screen with their own eyes, would it have remained the default?

Sometimes it will be important to correct the discrepancy between theory and practice, but usually not. You only need to understand it in a way that informs further action. You may need to keep these details away from executives, or risk being sent down irrelevant rabbit holes.

Mapmaking is social; Marina may hold the literal pen, but it grows in utility and accuracy only if people from every corner of the organization can weigh in on it. When new people come on the team or enter the room, encourage them to look and question its assumptions. Does that data sync really only happen once every twenty-four hours? Are those firewall rules even on?

This gets particularly interesting when two (or more!) people hold diametrically opposed views of the same piece of the puzzle.

Many times, these people never argued over the details before *because nobody knew the details*. They each had separate, somewhat vague ideas of how things worked that aligned with their perspective of the world, based on their role and past experience. We reminded you earlier that people hate cognitive dissonance and will do most anything to escape the feeling. Leverage this! If two people disagree about what happens at a certain process step, take them over to where it happens and let them see for themselves.

This works when both people operate outside the control loop in question, but can also be powerful when you need to change a specific control loop. For example, Marina once wanted to shorten the time it took the VA to process veteran disability claims, which at the time took literal years. At one step of her systems map, doctors wrote detailed medical descriptions of each veteran's situation, which took days. The doctors told Marina confidently that an even *more* experienced doctor then read their report and made a nuanced, thoughtful decision as to how debilitating the situation was. But Marina knew this was not true; at the next step, administrative personnel with no medical background at all strained to find the specific information they needed (such as how many degrees the veteran's left elbow bent) amid all the medical jargon, and often had to send the veteran back for another exam to get it.

She was never going to persuade the doctors individually to stop writing these essays, so she leveraged sensemaking instead: She invited the doctors and the administrative staff to meet and show one another their work. The doctors immediately lost their minds when they saw that their days of careful work were effectively worthless, and they refused to do it anymore, removing a huge source of wasted time from the process.

HAVE YOU TRIED TURNING IT OFF AND ON AGAIN?

Once you have a rough initial sketch of your system, we encourage you to test it out by asking an important question for understanding what controls exist and where you might find the operators:

If you want to stop the existing system completely, how do you do it?

Who could make such a decision? Whose fingers would touch what controls to make it happen? How would the order be communicated from one to the other?

You can expect these questions to make people uncomfortable, and the answers will probably be vague and speculative. If the system is internet connected or represents basic infrastructure like health care or a utility, the organization may have never intentionally shut it down before. Nonetheless, if you can pin down a few basic answers, it will help clarify the finer-grained questions of authority and capability that are coming. After all, the authority that can destroy a thing has sufficient power to make changes to it.[iv]

"How do you turn it off?" is a useful bedrock starting point because it has an answer in any system. It is also nearly universally true that a system must be stopped to safely make changes to its machine behaviors. The system you care about will likely have additional concepts of "on" and "off" that are more nuanced.

[iv] So says Paul Atreides (in *Dune*). Weird regulatory regimes with separation of duties can lead to exceptions, where a group with the power to turn a machine off may not have the power to modify its behavior.

NERDY SIDEBAR: DIFFERENCES BETWEEN BATCH-ORIENTED AND TRANSACTIONAL SYSTEMS

Many systems have a pipeline structure, where new input triggers a series of process steps of indeterminate length before ultimately changing state in a database of record or generating an output artifact such as a standard widget. Hospitals and factories work this way, as do banks and insurance companies.

It is best to stop such a system by blocking new input and waiting for all intermediate steps to "drain." The only other option is that if every process step gracefully resumes its input and output queues after an interruption, then you can stop the whole pipeline in arbitrary order. This idea is usually more aspirational than actually true. But if you have no insight into the intermediate steps, or if they take unacceptably long to drain, then you will have to take your chances.

A system that is not structured as a pipeline of batch tasks is probably "online transactional." By this we mean any system where user actions are processed and resolved while the user waits. Such systems are more tightly coupled than batch pipelines, for better and worse. The difficulty will be in determining what constitutes a "transaction" that the system owners care about. For instance, in a typical REST-style web application, textbooks consider a "transaction" to be an HTTP request and response, which may be resolved in milliseconds. The aspiration is that you can stop, start, move, and modify serving components in milliseconds, with your customers none the wiser. The problem is that the transaction the business cares about is something like "a customer submits a health care application," which will comprise thousands of REST transactions across minutes to hours. Thus you may find that in real life, the graceful shutdown of your stateless microtransactional hotness looks the same as the sad, hand-cranked COBOL batch processing mainframe next door.

Of course, the most common architecture is a hodgepodge of batch- and online-style services that all maintain their own fragile states and are monkey-patched together without any strategy. One agency's system for processing applications for unemployment benefits encompassed a DB2 database on a mainframe, a complex set of TN3270 ("green screen") interfaces, undocumented COBOL batch jobs, and another set of browser-based web applications. The web apps interacted with the DB2 database at arm's length, while they maintained their own state in whatever RDBMS, flat file, or spreadsheet that particular developer found convenient. As you might imagine, this caused inconsistencies and data loss, which were managed by caseworkers.

These caseworkers, coping with unreliable databases, managed their files by creating their own unconnected Excel spreadsheets, circulated by email. How you turn off such a thing is a question without a straightforward answer.

If this is your lot, you may have to find a creative way to lock users out. There is usually an intranet reverse proxy or firewall that can be disabled, or an inbound phone number that can be redirected.[v]

You can move on from the question of "how to turn it off" once you have roughed out reasonably plausible answers to who, when, and how. It's not a high-value exercise to actually do it just for the sake of proving you can. (A real reason to do it may turn up soon enough!) As we will see, the purpose of this exercise is mostly to force the organization into concrete answers about control that refine our map, our people, and our understanding of what we can do next.

---

v What actually happened in this case was that we eventually had to halt the acceptance of new applications entirely for a couple of weeks, while internal processes were rerouted. It was not possible to hide this from the public.

Specifically, what we are mapping here is the set of sysadmins, network operators, security guards, janitors, animal control officers, and any other person, key, or credential that is necessary to exercise basic operational control of the system. You should expect a lot of answers that sound like, "I guess if I was going to do that, I would have to call Dave and let him know." You should ask follow-up questions to find out who each Dave is and keep track of the answers.

WHEN TO STOP UPDATING THE MAP

Information-gathering and mapmaking can go on indefinitely, as sensemaking has no end, so watch for the point of diminishing returns. Use these heuristics to decide when it's time to move on and prioritize taking novel actions over refining the map:

- Are you hearing the same stories in every interview?
- Can you predict the content of your next meeting with reasonable accuracy?
- Can you name the databases or systems representing key processes?
- Do you understand the routine, nonexceptional workflows?
- Can you describe how data changes align with real-world actions?
- Are you fluent in the jargon used by the people who run the processes?

When you have a few "yes" answers to the above, you have enough credibility to bootstrap the sensemaking loop that you are going to construct. We usually reach this point in a few days, and can do it in less time when necessary.

MAPS ARE VALUABLE AFTER A CRISIS, TOO

A map is never really finished. Systems evolve, processes change, and new issues emerge over time. Treat the map as a permanent sense-making exercise. Make it routine to revisit and revise it—especially after incidents, major changes, or new discoveries. Regular updates can preserve the crisis engineering team's hard-earned knowledge and keep understanding current, reducing the risk of blind spots in the future.

If we were hanging around an organization after a crisis, we'd consider a plausible system map to be a treasure map of sorts for how to implement the next phase of changes and improvements we'd like to see, after the crisis ends. In a crisis, the organization is forced to reckon with reality, and this reality will soon become its new normal. You influence what that new normal looks like by strategically capturing and updating information on this map.

As you map your shared environment, those social interactions will help you find your crisis engineering team members, along with a myriad of external teammates who can provide input, explain intricacies, or move barriers. We'll explain how to find these people next.

CRISIS CHEAT SHEET

All existing models of the complex system you are trying to take actions on, be they mental or written, are wrong and out of date. Use the access and urgency created by the crisis circumstances around you, centered from a crisis engineering center, to cocreate a plausible map of how information actually flows through the system and what happens at each step.

Why This Matters

- You can't take actions on what you don't know exists or have no working understanding of.
- An inaccurate but plausible map is immediately useful for picking actions to take in order to observe outcomes and gain further information about the system.
- The social process of mapmaking is an effective tool for permanently changing minds, which is part of the leverage a crisis period provides for making outsized organizational changes.

The goal of mapping the system is to iteratively converge on a consensus reality: a single, shared understanding of how the system operates. The consensus reality should span process, people, and data. This sets the stage for objective measurements, action plans, and the ability to reimagine what is possible.

What to Map

- Control loops: Note *controllers* (something operating on internal rules), *sensors* (its inputs), and *actuators* (available

actions). Controllers can be machines or people running algorithms or exercising discretion.
- Zoom *in* when you are diagnosing problems you can do nothing about.
- Zoom *out* when a higher-level controller (like an executive or a department) is doing something harmful and you don't understand why.

How to Map

Have the incident lead or a delegate (preferably an extroverted one) be responsible for mapping, with a buddy to tag along if you can afford it. Build the map iteratively by interviewing people. *Mapmaking is a social process.*

- **Start Anywhere:** Do not agonize over the starting point; guessing is fine.
- **Consider All Perspectives:** Think about user experience, business process, organizational context, and technical systems.
- **Stick to the Basics:** Ask simple questions like: How does this work? Where is this system, physically? How many of this are there? How often does this change? Who knows more about this component or process?
- **Observe Directly:** Talk to people and look at dashboards or policies firsthand; do not rely on existing documentation.
- **Ask for Mysteries:** Literally ask people to tell you about the parts of the system that remain a mystery to them.
- **Let It Be Messy:** Plausible maps have some contradictions.
- **Mix It Up:** Intentionally ask two or three people in a team the same questions separately.
- **Assume Falsehood:** All answers are lies unless you observe them with your own eyes in a live system.

Stop mapping when additional interviews are producing decreasing returns, even though the map will be incomplete. Initial mapmaking should not exceed a couple of days.

Maps can be used to change minds.

Maps can be tested by using them to find an answer to a single question: How would we stop the existing system completely?

2.4

Find Your People

> One does not live for months or years in a particular position in an organization, exposed to some streams of communication, shielded from others, without the most profound effects upon what one knows, believes, attends to, hopes, wishes, emphasizes, fears, and proposes.
>
> —HERBERT A. SIMON, *ADMINISTRATIVE BEHAVIOR*

Sensemaking is social. While mapping the system, you'll also need to find and recruit members to the crisis engineering team. Mapping, it turns out, is one of the best ways to find those teammates. Talking to all the operators instead of reading documentation will give you a more useful picture of what's actually happening, and will connect you to people who can help you decipher mysteries and referee disputes on ground truth.

For the crisis engineering center, your goal is to find the smallest set of people such that for any action you might need to take, the decision can be made and implemented by the people in the room. There seems to be no minimum size: If two people form a sufficient set, start with them. There does seem to be a maximum size: If your list grows past ten to fifteen people (the famous "two-pizza team"), narrow it down. People often think this constraint can't possibly be true for their own organization, because it's so big and complicated! Take comfort and confidence in our experience: We have seen groups of about ten people solve problems in the largest, most complex organizations in the world.

With this set of people, you are going to disrupt their individual existing working patterns and establish a new culture and cadence. You can improve the chances of success by taking care with the choice of individuals, as well as the overall team composition.

The composition of this team will shift as the center of focus shifts. This is good. The interactions you have across the organization (and sometimes even outside of it) can build a deep bench from which you can pull future team members as necessary.

Even if someone never joins the crisis team, they may still have a valuable role to play. A crisis engineer builds a robust Rolodex of diverse identities from every angle of the organization who can quickly (we're talking seconds) give feedback on proposed actions, report back on results, and further fix your shared reality (your map).

CHARACTERISTICS OF IDEAL CRISIS ENGINEERING TEAM MEMBERS

You want individuals with an action bias: They prefer doing something to doing nothing.

They should not be at either end of the spectrum of loyalty to the status quo. You will not get useful contributions from a person who is constantly brainstorming reasons why a change won't work. Neither will you be able to use a person who is excessively critical—they won't be able to resist using the disruptive process to push their standing agenda, and the organization will be highly sensitized and resistant to them.

The individuals you choose should have some unique skill or area of responsibility that makes it clear they have a role to play. They don't need to know every last detail—only enough that their guesses are going to be correct more often than not.

They need to be strong enough communicators that you don't have to spend energy managing their interactions with other team members.

You are working toward a team that can cover all the critical systems in terms of access and permissions. It's easier, but not strictly necessary, if the members are already acquainted with one another. On the other hand, it's a bad sign if you collect people who already work together all the time. If your organization functions *at all*, it already has informal channels to get things done. If you simply re-create this existing group, you'll have the same gaps in insight you have today.

People we look for include:

- The communications team's go-to person for explaining how things work in plain language
- The person mentioned as an expert during seemingly disparate engineering discussions
- Someone who disregards formal organizational boundaries to get things done
- The quiet developer who repeatedly contributes surprising insights about edge cases of systems administration
- Someone with no orthodoxy about which technology or programming language to use
- Someone who will make a pragmatic decision based on the realities of the environment, not based on a pie-in-the-sky ideal world
- Someone who will happily chase a bug from an obscure user report through multiple systems (and organizational layers) to reach a resolution
- Someone with hobbies or past careers that require rapid decision-making and prioritization—think theater, live music, event production, the military, or emergency response of any kind

If multiple people refer to someone as "that one person that we go to when we need something from department X," you are on the right track.

A favorite example of an unconventional but highly successful team member:

> In the beef packing business I found that Sinclair was of real use. I have an utter contempt for him. He is hysterical, unbalanced, and untruthful. Three-fourths of the things he said were absolute falsehoods. For some of the remainder there was only a basis of truth. Nevertheless, in this particular crisis he was of service to us, and yet I had to explain again and again to well-meaning people that I could not afford to disregard ugly things that had been found out simply because I did not like the man who had helped in finding them out.
>
> —President Theodore Roosevelt on Upton Sinclair[1]

As you read this list, it's worth thinking about which of these characteristics *you* have right now, and which you might want to develop to be of use in a future crisis.

HOW TO FIND THESE TEAM MEMBERS

You will often find these individuals while mapping the system, as described in the previous chapter. Mapping should take you to the system operators, programmers, lawyers, and others who have the answers, reach, and action bias you need.

If we did not persuade you earlier that it's absolutely critical to follow processes and see systems firsthand, and not to rely on reports or manuals or screenshots, we hope that discovering the people you need to actually solve your crisis along the way gives you another reason to do it. Go to the source. Go to the basement. Marina has had notable success going to the restroom, where people confide in her or slide confidential reports into her purse (which she now "accidentally" leaves in a conveniently accessible place).

When you meet someone with strong knowledge of a business process, ask them if they have an equivalent contact on the tech side. Often, two useful people find each other and keep an open line to get things done under the radar.

As you conduct inquiries, constantly ask "Who else should I talk to?" or "Who might know the answer?" Again, not everyone will be part of the core team, but many people you find will hold pieces of the puzzle.

Find existing coordination channels; these are likely not labeled. There may be a monthly or quarterly phone call or even happy hour where people from different corners of your organization come together to talk and exchange notes. Leveraging this channel can fast-track your crisis engineering center.

WHAT TO DO WHEN YOU FIND TEAM MEMBERS

When you find someone who seems like a good fit for the team, invite them into the crisis engineering center. Do your best to succinctly explain how the crisis engineering team works and the unique value you expect this person to bring. Bluntly point out that there's not a lot of time to cajole participants or give them much time to consider the offer. We've had success framing the invitation as: There is a crisis, this team has been empowered to fix it, the work is going to be really hard, and you don't know much about what's going to happen next—but you will find out together. If someone protests too much and/or requires too much certainty about the future, quickly move on to someone else.

Once they agree to join up, publicly introduce them to all current team members, explicitly outlining their respective strengths and roles. This helps to break down silos and encourages faster collaboration; you don't have time for them to storm and norm.[i] When

---

i   Team building researchers refer to the initial stages of a new team as "storming" (working out conflict) and "norming" (developing a new, peaceful working order postconflict).

possible, use tools like shared documentation, group chats, and regular stand-up meetings to keep everyone in the loop and make it possible for a new team member to quickly get up to speed. A shared understanding of goals and progress is how you get to the shared sense of reality you need for productive sensemaking.

Beware of accidentally importing a preexisting clique. If you are working with six people, and three of them are on a bowling team together, you will have to take extra care to integrate the nonbowlers. How you do this will depend on local cultural context.

When someone's contributions become less frequent or useful, let them leave in success. As the understanding of the issue changes, so will the composition of the team.

There's nothing shameful about a person graduating out of the team because the issues in their area are resolved or not relevant. *This is behavior to normalize.*

WHAT TO DO WHEN YOU FIND HELPERS OUTSIDE THE TEAM

In your travels through the system, you will likely encounter many people who are experts in very narrow topics. They don't need to be part of the core team, since they probably need to focus on propping up their part of the system. They should be noted as a point of contact for their specialty and marked on the system map.

For example, we commonly find that across a system, there are undocumented codes (such as errors or order status), and no one person knows what they all mean. However, *someone* knows what each of those codes means. You may have to consult many people to assemble an authoritative list.

Find Resilience

Human operators will be needed to resolve almost all failures. Practitioners and first-line management make adaptations constantly to maximize production and minimize accidents.[ii]

For any given subsystem or component, there exists a person who knows it better than anyone else. They are rarely in a position of authority, yet their insight is indispensable. Identify this person, listen to them, and ask them with whom to speak next. As you gather stories, come back to them with your findings and refine your map.

The truth doesn't have to be all doom and gloom; there are always reservoirs of resilience. These are people and systems that are left off the official diagram but are load-bearing pillars that keep everything working, whether it be Betsy running manual SQL queries to make a report work, or an automated script that clears out system alerts so the cache doesn't overflow.

Bespoke tools, "temporary" workarounds that are decades old, and ad hoc scripts can be viewed as inadequate or in need of replacing, but in every large system we've ever seen, they are essential parts of what's going *right*. Discovering them is not a cue to shut them down or bury them in a modernization project. Instead, consider them a potential source for the flexibility you need. The people behind these workarounds are signposts of the complex system's adaptability and resilience.

Elevate the Voices of Those with High-Quality Information

It is our experience that the people with the best answers and ideas have not historically been asked to share them. Crisis engineering requires primary sources. When you get a secondhand report, invite

---

[ii] "Human practitioners are the adaptable element of complex systems" (Richard I. Cook, "How Complex Systems Fail," https://how.complexsystems.fail/).

the original source into the room. Only share on behalf of others when the source has legitimate concerns about confidentiality or there are interpersonal dynamics at play (such as not being able to publicly disagree with their boss).

Ask What They Need, and Get It for Them

Ask the people who help you directly what tools, access, or support they require to perform effectively. Whether it's approval for additional resources, a software license, or permission to take action, your responsiveness to their needs signals you value their work and are committed to enabling their success. This also helps remove potential bottlenecks before they escalate and prolong or worsen the crisis.

Some of the most seemingly simple requests we've ever granted—such as chairs, power cords, and for the air-conditioning to run on weekends—made material impacts in the ability of the requesters to do their job and implement solutions.

Protect Your Sources

Crisis situations often bring to light uncomfortable truths, systemic vulnerabilities, or other strongly negative information. The people who share these insights can feel exposed or at risk. Take care to shield their identities and ensure they do not suffer negative consequences for their candor. For example, anonymize sensitive feedback when reporting findings, and actively counter any attempts to scapegoat or discredit these contributors. By protecting your sources, you can encourage a continued flow of honest and actionable information. Whenever possible, publicly praise truth tellers to demonstrate that it is a valued behavior.

Highlight Contributions

Recognition matters. Celebrate the achievements of contributors, no matter how small they might seem in the broader scope of the crisis. A shout-out in a meeting, an email to a senior leader or manager, or a message in a shared chat can go a long way toward keeping morale high. When people see their efforts are valued, they're more likely to stay committed and enthusiastic. Help them see themselves as important parts of the new reality.

If you miss a chance to recognize someone in real time, keep a list for when you have a moment after the crisis.

WHAT ABOUT EVERYONE ELSE?

Not everyone in the organization will be an active participant, or even an ally, in your crisis engineering efforts. Some might be skeptical, obstructive, or simply indifferent. There are a few important things to consider in your approach to these individuals.

Understand Their Relationships

Mapping out the informal networks and alliances within your organization is critical, as we discussed in the prior chapter. Who listens to whom? Who influences decisions, with or without formal authority? By understanding these dynamics, you can anticipate resistance and plan your approach accordingly. Sometimes, winning over a key influencer can turn a vocal critic into a cautious ally.

> We once worked to understand why a state IT modernization project was stalled. As usual, there were multiple government agencies and multiple contractors. It was clear that part of the problem was some of these people didn't like each other. Yet after dozens of interviews, we were utterly unable to get anyone to explain because no one wanted to bad-mouth anyone. Then we went to the company hot dog picnic, which featured a dunk tank as a fundraiser. All became clear when we saw which employees lined up for the chance to dunk which executives.

Understand Their Risk and Incentive Frameworks

In the course of ending the crisis—not to mention making lasting change—you will likely need many people outside the team to do work, make decisions, and accept responsibility they are not particularly excited to take on. People resist change for a reason. Often, this reason stems from their perception of risk. What are they afraid of losing? Their job security? Their next promotion? Their reputation? Their control over a process? Their experience and seniority with the status quo?

People's actions are also shaped by what they're rewarded for. Understand the metrics or goals driving the behavior of anyone whose participation is required in order to take necessary actions. Are they evaluated based on quarterly results? Operational stability? Compliance metrics? Obscuring or silencing problems?

By identifying how they see themselves within the world, you can come up with ways to mitigate their perception of risk or better align with their goals.[iii] For example, you could tailor your communication to show how the crisis response aligns with or supports their

---

[iii] We recommend reading about conservers, climbers, and zealots in *Inside Bureaucracy* by Anthony Downs.

incentives. When possible, create opportunities for them to share in the success of your efforts. Later, when talking about ending a crisis engineering effort, we'll discuss how to preserve this shared identity across the team.

Share Information When Possible

As you work through the crisis, the network you've cultivated needs to remain engaged. Regular updates, even informal ones, help ensure that the people you've brought together stay connected and aligned. These updates don't need to be exhaustive; a succinct message summarizing recent progress, current challenges, and next steps is sufficient. This practice fosters trust and ensures the team's focus remains sharp.

Whether supporters or naysayers, when the crisis response is a black box, people will make up—and spread—their own stories. You can help mitigate this by ensuring others have a way to get accurate, timely communications about the crisis. This will help most people have a shared story about what happened, and what is going on.

We'll talk more about sharing news and updates in chapter 2.8 (Communicate in a Crisis), but for now, remember that the way you interact now can permanently change the connections and norms in the organization. This is your chance to cocreate a new reality. If you cocreate this reality only in the center, you may emerge in an alternate universe.

Navigating the complexities of assembling and managing a crisis team requires not just technical expertise but also emotional intelligence. By staying in touch, connecting the team, normalizing a shifting team makeup with the cadence of the crisis, and understanding the broader organizational dynamics, you can ensure that your efforts not only address the crisis at hand but also build a foundation for stronger, more resilient systems moving forward.

CRISIS CHEAT SHEET

Assemble a small team (two to ten people) who can, among themselves, decide and implement changes *right now*.

Who are you looking for on the crisis engineering team?

- Members must have an *action bias*, preferring to do something instead of nothing.
- Avoid people who are either extremely loyal or disloyal to the organization or status quo.
- Members should have some unique skill or area of responsibility, capable of guessing correctly more often than not.
- Members need to have good communications skills so that you don't spend your time managing their interactions with others.

How do you find them?

- You will mostly find the right people while mapping the system.
- Go to the source, go to places where people informally or formally congregate (e.g., basement, cafeteria, happy hour).
- Ask the business who they rely on for answers or solving problems.
- Ask everyone you find: "Who else should I talk to?" or "Who might know?"
- Find existing coordination channels.

What do you do when you find them?

- Bring them into the crisis engineering center and publicly introduce them to everyone there, and explain why they are there.
- Avoid importing a preexisting clique.
- Normalize graceful exits as "graduations" as members become less useful or relevant to the crisis engineering effort over time.

What about the other helpful people you find?

- Most people you meet won't join the team, but can still answer questions, validate ideas, and explain obscure subsystems.
- Maintain a list of these contacts to lean on as the crisis effort continues.
- Ask them what they need and get it for them.
- Elevate their voices if they have high-quality information.
- Publicly praise truth tellers when possible, and always protect your sources if they provide sensitive feedback.
- Highlight contributions; keep a list if you don't have time to praise in the moment.

What about everyone else?

- Understand the relationships of skeptical, obtrusive, or indifferent people.
- Understand the risk and incentive frameworks of people.
- Share information with the network you've cultivated with your map.

2.5

Take Novel Actions

> When asked whether he had "ever been instructed in setting an escape fire," Dodge replied, "Not that I know of. It just seemed the logical thing to do."
>
> —NORMAN MACLEAN, *YOUNG MEN AND FIRE*

We chose to dedicate this book to a particularly memorable novel action: Wag Dodge's escape fire. Facing certain death, Wag—the foreman of the smoke jumper team deployed to put out the Mann Gulch fire—set a *second* fire, jumped inside its circle, and ordered his men to do the same. Thinking his action crazy, the others continued running away from the blaze. They perished, while Wag survived. His novel action saved his life, and the lives of many future firefighters.[i]

Picking up a marker to map out your system and finding the operators across the organization is only the beginning of an effective sensemaking process. It is then necessary to take action on the evolving system to deepen your understanding of the current state and build the path out of crisis. This is the retrospective part of sensemaking: not looking *at* a map, but rather looking *after* taking action to make sense of what just happened.

---

i An escape fire creates a safe space within a larger fire by proactively burning all combustible material, leaving no place for an advancing fire to take hold.

Even a crude and inaccurate map is enough to suggest actions that are more likely than not to be productive, such as adding server capacity or monitoring, or suspending a troublesome practice or process. The map will improve rapidly as you observe the results of those actions. With a good enough map and a wide enough scope of control, any amount of dysfunction can be resolved.

OPENING MOVES

While the best first novel action can vary, we have one that we start with more often than not: Can you change one word in one line of code? And if you can successfully do that, what other seemingly routine operation, like performing some system maintenance chores, can you do next to reveal how the system actually works?

Can You Make a One-Line Code Change?

One blunt question is usually enough to start taking action on your system.

Earlier, we mapped how to turn the system off and on again. Now it's time to take action on that system by making a one-line code change and see how the map stands up. No matter what your problem is, we are willing to bet there's some computer in the way of solving it.

Try this exercise right now: Change one word on a screen of an affected system.

The learning opportunities will include: Who decides how to modify system behavior? Whose fingers touch the keyboards that make changes to the source code? What country and time zone are they in, and what language do they speak?[ii] Where and how does

---

[ii] Implementation of computer code is often outsourced to a low-priced team overseas. We have yet to find an organization that contracts out product or design decisions in this way.

that person receive instructions? What language is the source code written in? Where is the code repository[iii] (if there is one) and who has access? What approvals are required at what step? What change management processes have to be satisfied?

Upper management will get impatient with such basic questions. They believe the answers are obvious, known by everyone, and not relevant to the real problem. You will need to commiserate with these feelings but persist nonetheless, explaining (accurately) that the crisis engineering effort can't possibly help with those real problems if it can't make a one-line code change.

When you get past this problem, you may next encounter panic from the executives who have just learned that the "obvious answers known by everyone" were wrong.

Enumerate More Routine Machine Operations

If you reach the point where you can start and stop the machine at will, and you can make code changes, there is nothing about the machine behavior that you can't fix. You should have a fairly good picture of what kinds of improvements are possible in the time that you have. (If you don't, see the section "Ideas for Novel Actions" on page 166.) Carrying on with our analogy, you should be able to draw a map that shows you one path to success, and renders rough time-and-effort estimates.

If time permits, it's worth following up on some loose threads and asking a few extra questions to find any unusual hazards or shortcuts in other parts of your map.

Asking questions like "how to turn it off" earlier gave you reason to meet some of the operational staff. Do your best to find out about all the routine maintenance they manage. Usually, there is a collection of tasks such as deploying new code, deleting expired tickets, archiving logs, backing up a database, clearing records representing

---

iii A **repository** is a central place to store code where everyone working on it can see the code, see one another's changes, and see the history of changes.

discharged patients, and a half dozen other things you have never heard of. They will be in varying states of automation, and their timing and cadence will be somewhere between "scheduled, scripted task" and "randomly."

Again, it will not be obvious to some people why you are talking to a data center tech from the overnight shift, in detail, about how she tests the batteries in the racks on the Friday after each full moon. Yet we have almost always found it will reveal truths such as:

- When do scheduled service disruptions happen?
- How well does management tolerate extended outages or operational mistakes?
- Where are procedures documented (if anywhere)? How big is the space between documentation and practice?
- Do operators know any shortcuts you can repurpose now?
- What parts of the machine system are unusually fragile? What is everyone afraid to touch?

Some large systems have scattered operational responsibilities far and wide, across many teams and many contractors. In such a situation, you will have to triage. You will not have time to talk to everyone, but you should always talk to the people who control the core functions and known trouble spots.[iv] It will be impossible to find every manual process, but the core system operators will be happy to tell you about weak points outside of their area of control.

[iv] If your crisis involves a web application, we can give you this list now: the load-balancing apparatus, the customer-facing web servers, and the database.

> One morning in the middle of the HealthCare.gov rescue, we discovered a widespread performance problem. Across the site, machines performing completely different tasks were all acting as if they had lost half their resources overnight. The distribution of affected machines did not fit any known model, such as a bad code change (which should affect only one type of application server) or a data center infrastructure brownout (which should affect only one site). Then, from one of the desks in the back row, someone offered up the critical clue: "Steve left on vacation today."
>
> This led us to the root cause: The agency required a particular enterprise antivirus program on every machine. The antivirus ran a full-disk scan every day, right at the beginning of business hours, on all machines. To stop this strain on the system, Steve manually turned off the antivirus scan every morning. As you might imagine, this adaptation was not documented anywhere. We would never have known about it if Steve had not gone on vacation, or if no one in the room noticed his absence.

TAKE EASY ROADS

We believe a crisis is a small window in time when your efforts are naturally amplified by the environment, rather than naturally resisted. It's also true that at any moment in time, there are actions that will be naturally easy and actions that will be naturally hard. We exploit easy ones as much as we can. In workshops we describe this as "rolling boulders downhill instead of uphill."

It's worth asking why rolling boulders downhill isn't more popular. A little of it is because of the bag of vaguely Puritan values that is deeply ingrained in our particular culture: Virtue is hard work, hard work is virtue, hard work is pain, therefore virtue is pain. We authors have given up on causes and organizations that confused working hard with being effective.

There are usually many ways to get something done: Look for the actions that require the least amount of struggle.

Obvious enough. And yet, most systems have enough roll-ready boulders to keep a small team busy for a while. They may have been overlooked, or they may have only become roll-ready in the latest crisis. Here are a few of our general strategies for finding them.

USE HUMANS AND MACHINES EFFECTIVELY

This sounds obvious. But the current incoherent debate over AI suggests that many people don't realize humans and machines have different capabilities, strengths, and weaknesses. To wit:

Machines
- Are reliable
- Never get tired
- Follow instructions precisely (including "bad" instructions)
- Do arithmetic fast
- Can be fully controlled
- Cannot handle unanticipated situations

People
- Are not very reliable
- Can focus for only short periods
- Can work with incomplete instructions
- Do arithmetic slowly, with mistakes
- Cannot be fully controlled
- Can handle unanticipated situations

The human strengths around ambiguity are enormous. The success of a complex system is determined by its **adaptive capacity**,[1] and the adaptive capacity is supplied by humans.[2]

As a matter of good systems design, you should automate all the rote work, and route everything that is ambiguous and requires judgment to a person. Unfortunately, this is not how things actually work.

Even profit-making corporations employ legions of people doing tasks that can be automated. Call centers and bank tellers perform adaptive functions that are indispensable, yet the majority of their time goes to routine transactions that could have been done online or at an ATM. For some companies, it is a matter of providing good service. On the other hand, such as when an insurance company does it, you can bet it's because they have calculated that it's cheaper to have a call center than not.

Governments are even more likely to do it. Most large programs operate with a certain amount of gravy to be ladled out in the form of big contracts for call centers and grants to community-based organizations for "assisting." Do retail agents help customers get better outcomes? Sure. These jobs also certainly help get the programs off the ground. In the coalition building before the policy exists, there will be much talk about the number of jobs created. After the programs are in operation, nobody looks very closely at facts like how many calls per day are really coming in. No good can come of that. New recruits to the U.S. Digital Service all discovered sooner or later that every federal program comes with an unspoken secondary goal of being a jobs program.

Whatever the reasons, you will find rote tasks that should be done by machines that are misallocated to humans. You won't be the first person to notice. You will quickly be told the list of reasons why this cannot change. But where others have given up, you may now have an opportunity, given the crisis at hand.

This is where backlogs are useful. You wouldn't be here asking questions if there wasn't a problem with system performance *somewhere*. You can overcome some fears about job elimination while there are large piles of work sitting undone, but you have to find the piles first (hence, mapping your system).

The second way that organizations work against the efficient allocation of humans and machines is that they are powerfully motivated to create **accountability sinks**. As described by Dan Davies in *The Unaccountability Machine*, the ideal process to a bureaucracy is one in which no decision can be attributed to any person. On HealthCare.gov, remember, carefully crafted contracts kept any one contractor from being held responsible for the overall failure of the project. Identifiable decision-makers might be required to answer questions from a regulator, which is bad, or might be found to have done something wrong, which is worse. Thus, you find tasks that require judgment and discretion delegated to a machine that *cannot actually handle them*.

It's important to understand that the individuals who work this way aren't wrong, in the context of the system in which they live. If they were wrong, you could replace them with better people and get better results.[v]

Your system will have other opportunities for improvement without struggle that we haven't anticipated. The generally useful skill you can practice is to value and prioritize these opportunities. Resist the management impulse to distribute rewards based on how hard people work, rather than their results. Sometimes it won't seem fair, and it will require extra work to keep things running smoothly. But the same number of people working the same number of hours will get more done under this mindset.

---

v   This doesn't work. We checked.

EARNING BUY-IN FOR RAPID NOVEL ACTIONS

Ending a crisis is often a struggle against rapid change. Organizations tend to dig in their heels, insisting that more of the same (e.g., more overtime, more humans working the same process) is the only way out—even though it's our lived experience that this never works. Balancing innovation and risk is essential.

Change Existing Tasks Instead of Creating New Tasks: In *Bureaucracy*, James Q. Wilson argues that employees respond well to change when their task remains "the same" but are challenged when their tasks shift. Does your novel action ask an employee to still answer the phone and complete a form, only it's a slightly different form? Or are you asking them to conduct their work completely differently? This will matter a lot when it comes to getting people to change behaviors quickly.

Pilot Whenever Possible: Test new processes or changes on a small scale before rolling them out broadly. This minimizes risk and allows for quick adjustments. For example, one team or desk area, or one regional office, might try something for a day before it rolls out any further.

Get Feedback: Involve those impacted by changes in the decision-making process. Their insights can help you avoid unintended consequences.

Balance Effort and Reward: Secure support from key stakeholders by trading efforts that require employees to perform new or novel work against requests to remove tedious, unwanted tasks from their plate. It's often possible to do both!

IDEAS FOR NOVEL ACTIONS

Need an idea for a novel action to take? Examples include:

- **Turn It Off.** If at all possible, even if the political stakes are high, stop new incoming orders or requests until you can get your arms around what's currently in the system.
- **Try a Process a New Way.** Maybe you can forego or streamline a policy requirement in a way that makes a material difference in processing time. This is the sort of change that might be a candidate for a permanent change postcrisis, especially if a temporary pause can prove it was not adding value or that removing it did not add or cause harm.
- **Eliminate or Change Steps.** Review your map and look for steps that are not adding value, such as mailing information packets, making manual reminder calls, or other steps that could be stopped without impact. These are also candidates for permanent elimination postcrisis. You could also consider modifying a step, like changing the medium (e.g., sending a required notice via email instead of physical mail in order to do it faster and with fewer humans licking envelopes).
- **Add or Remove Friction.**[vi] Make it easier to do a step you *want* to encourage by removing some of the friction around completing it. Or try the opposite: Discourage people from taking an action by intentionally making it more difficult.
- **Batch-Process Certain Items,** such as automatically approving (or denying) certain requests based on specific criteria, clearing informational alerts, or sending reminder messages about overdue tasks to external

---

vi For examples of both, we strongly recommend Robert I. Sutton and Huggy Rao, *The Friction Project* (St. Martin's Press, 2024).

users. (Be sure the task is actually overdue and not sitting in your backlog, though!)
- **Move People Around.** What happens if you move (qualified) people from later in the process to earlier steps? What if you allow less experienced employees to attempt certain work items? It may not always be possible but is worth thinking about, or even trying with a few people or for a short time.
- **Add Automation.** What steps are the most time-consuming to work down? Is there any form of automation that could partially or fully replace this manual work for some or all cases?
- **Rewrite in Plain Language.** If users are taking actions that cause more work for your employees, such as submitting the wrong form or consistently choosing the wrong option from a menu, try to push plain language instructions to users to prevent this issue going forward.
- **Enable Self-Service.** Can you jigger a self-service claims status tracker on a website and/or through an automated phone line that allows people to check their claim status on their own? Make sure the answer they get is in plain language, too—knowing your order is in "status F24" is not helpful, but knowing "it shipped Monday" *is*.
- **Unleash Pending Actions.** Especially if your organization is in the midst of a "modernization" effort, it's likely that many simple requests and pending changes have been tabled until the new system is in place. Implement them now, instead.
- **Expand People's Scope of Action.** In health care, this is known as practicing at the top of your license. Can you give the people you have now greater scopes of authority or decision-making?
- **Bring Back Retired Employees or Other Recently Reassigned Staff.** Hiring green staff is almost never the

answer to a problem, because training them takes all the attention of your most experienced, most needed staff members. But even a handful of experienced employees who can come back temporarily can sometimes make a dent.

Sometimes it's easiest to come up with a novel action when you're stuck on taking *any* action. Layer Aleph once took a team trip to Nauru, drawn by its status as the least visited country in the world. This is in part due to its remote location, but as your authors, who have been to every country in Oceania, can attest, it's actually much easier to get to Nauru than some of its neighbors. This designation has much more to do with its visa application process. We were prepared to submit full criminal background checks and X-rays of our lungs, but were caught off guard when the official Nauru email address[vii] for requesting a visa immediately bounced back with a "mailbox full" message.

Stuck, we tried a novel action: We looked for someone on social media who had recently traveled to Nauru. We commented on a YouTube video of someone boarding a Nauru Airlines flight; they helpfully wrote back that *of course* the way to get a visa is not the official email, but rather by contacting a completely different individual. We tried this advice, and some amount of paperwork later, we were in Nauru.

Would we have ever guessed that a YouTube comment was the only way to get a visa to a foreign country? No, but sense was made by taking action.

[vii] This is, literally, an @gmail.com address, making the "mailbox full" message all the wilder.

STOPPING ACTIONS

Is there anything more novel than *removing* something in an organization? As you build your map, you will likely start to see fully redundant or useless steps that you may want to stop. Common examples of actions we recommend stopping include overly burdensome approval chains (especially when they're serial, not parallel) and "status update" meetings.

We've found many cases where Step 2 collects data that is never saved in Step 3, only for that data to be re-collected at Step 4. While this duplication likely isn't a cause of your crisis, you are probably surrounded by individuals who are seriously overworked, and of whom this crisis is asking even more. If possible, this is a great time to permanently fix this inefficiency and give everyone a little more breathing room.

At the same time, it's easy to be overzealous in a crisis and try to slash away the entire forest of inefficiency. Organizational defenses may snap to attention if you try to slash too many things too quickly, even in the midst of a dire crisis.

USEFUL TOOLS FOR TAKING ACTION

> A complex system that works is invariably found to have evolved from a simple system that worked.
>
> —JOHN GALL, *THE SYSTEMS BIBLE*

By this point, you should have a list of hypotheses to test and the names of the people who can try them out. Now, we're going to look at some helpful tools when taking novel action.

We have spent a lot of words so far to convince you that sensemaking is the goal of a crisis engineering effort, that it requires novelty and action to progress, and that speed matters.

The most useful tools for taking action are not complex. In any mature organization there will be some reasonably well-understood and predictable tools already lying around. They'll be necessary for all sorts of operator practices that are required to keep the complex system functioning day-to-day. Perhaps there is some natural law that guarantees ad hoc workarounds and controls will spring up around human-built systems of sufficient complexity. On a manufacturing line, the equivalents are lengths of baling wire, metal shims, pry bars, hammers, and fuses. In high-tech systems, they include what Jennifer Mace calls **generic mitigations**:[3] already-built software and processes that perform functions that are useful in most or all crisis circumstances for limiting damage or isolating components. They will be tools that can be used before anyone has a full understanding of a problem. They will be made of "boring" technology, languages, and platforms familiar to a majority of practitioners—not the newest, shiniest technical fads.

If your complex system has a mainframe (or two, or ten) buried at its heart, great. It's not the enemy, it's probably fine, and now is probably the time to use it *more*. The older systems are more predictable and likely more reliable than the newer ones. We've seen a lot of mainframes; it sure looks to us like most of Western Civilization runs on them. They are not inherently a problem.

Many of the things a crisis center will decide to do in a complex system will fit into a few categories:

1. Changing the load, volume, or source of inbound events
2. Modifying or hijacking input(s) or output(s)
3. Modifying or replacing a component or subsystem entirely

We'll give some examples of the sorts of tools we've repeatedly found useful for these actions, and the sorts of ways we've introduced new ad hoc components during a crisis.

Tools for Changing Load on Components

- **Load Balancing/Traffic Routing:** Load balancing redistributes traffic across multiple components to avoid overloading any single one. Traffic routing policies can dynamically direct users to less burdened servers, regions, or resources. In a crisis, basic load balancing is a start, and you can tweak the configuration based on the evolving situation.
- **Blocklists/Allowlists:** A simple and powerful method to control input is defining what is allowed (or denied) access. For example, blocklists can temporarily exclude problematic users or regions that are causing excess load, while allowlists can prioritize critical users or systems that need uninterrupted service.
- **Degraded Operations:** When full service isn't sustainable, offering limited or core functionality helps preserve essential operations. For instance, an e-commerce site might temporarily disable image loading or advanced search features to reduce system load during high demand.
- **Planned Downtime:** Scheduling downtime to address performance or operational issues allows teams to implement fixes or upgrades without unexpected disruption. Communicate this to stakeholders upfront and plan it during off-peak hours, if you can.

Tools for Modifying/Hijacking Input or Output of Components

- **Pluggable Pipelines/ETL:** Extract-Transform-Load (ETL) tools and pluggable pipelines are useful for intercepting, modifying, or redirecting data as it flows through the system. For instance, rerouting malformed data to a

"quarantine" queue can prevent it from crashing downstream processes while keeping operations running.
- **Batch Processing (Mainframes) via CSV:** When newer systems falter, batch processing often remains a reliable fallback. For example, exporting data to CSV (comma separated values) for offline processing or analysis can bypass system bottlenecks, enabling teams to manually intervene or adjust outputs as needed. Just about any system can spit out and suck back in a CSV; this is a technique we use more often than not in a crisis.

Tools for Modifying Components

- **Fast/No-Gates Release Process:** Crises often require immediate changes. Streamlining approvals for emergency fixes or pre-authorizing specific actions can reduce delays. These are the sorts of exceptions or process reductions your decision-making authority can temporarily impose.
- **Static Content:** Swapping dynamic components for static ones reduces downstream load on other system components and the database(s). For instance, replace a personalized or dynamic home page with a static one, or replace dynamically generated public pages with static ones generated by regular batch processes.
- **Strangler Pattern:** This strategy (named by Martin Fowler after the "strangler" fig, or banyan tree, not a form of murder) incrementally replaces components of a system. Common candidates we've seen successfully "strangled" in a crisis include a claim or system status tracker, a new application or order form, or a new identity verification pathway. In a crisis, you might divert specific workflows to simpler, more reliable replacements, gradually deprecating the troubled system.

- **Turning Off Features:** Disabling noncritical features can drastically reduce complexity and load. For instance, turning off a recommendation engine or some analytics temporarily lightens system demands so you can focus on core operations.

Tools for Quickly Creating New Components

- **Spreadsheets:** Often underestimated, spreadsheets offer a fast and flexible way to collect, process, and visualize data. They can act as temporary databases or dashboards, enabling quick analysis and decisions during a crisis.
- **Scripting Languages:** Scripts are a lightweight and fast solution for automating tasks, running queries, or creating temporary workarounds. Python or shell scripts can quickly address immediate needs without waiting for a full engineering effort. You likely have many of these hanging around anyway, performing core bridging functions. It is likely that these sorts of components can be deployed more quickly than other kinds of software changes.
    - In particular, look at automating certain laborious tasks in the front-end interface. A scripting tool can automatically click around the screen and input information like a human would. Prioritize the most time-consuming or error-prone tasks first by pairing an experienced employee with a skilled scriptwriter. This can be used to confirm current features haven't stopped working, or even to perform certain bulk actions (like checking a box or clicking a button) across many screens so humans don't have to.

- **Regularly Scheduled, Automated "Cron" Jobs:** Automating repetitive tasks with cron jobs in a single, centralized location can maintain system stability by running maintenance scripts, monitoring processes, or triggering alerts. Once automated, they free up humans for other work. For example:

- **Run a Script**—This may generate a count of work items, archive or delete logs that are taking up space, or clear nonessential, informational-only (or severely outdated) work items from the queues of overly stressed agents.
- **Run a Batch Job**—Take the script you ran, and run it on a schedule. This can be used to automate some processes, reassign work items, automate certain messages, or automatically approve or otherwise move forward a claim.

Paradoxically, the less time you think you have, the more actions you need to take, and at a faster speed. If planes are grounded or surgeries are on hold, you can't spare the time to get too many opinions or compare too many alternatives. If you have a horizon measuring in days or weeks, you still need to take action, but can space them out slightly more.

Remember! Every time an action is taken, the incident lead must write it down in the crisis journal with a date and time stamp.

WHO SHOULD TAKE THE ACTION?

Sometimes, it's clear who needs to take an action—the administrator for Database A pulls the report, the assigned lawyer updates the company policy, or someone on Amy's team pushes the code change.

Other times, it's less clear. In that situation, start with a member of the crisis engineering team. If no one in the room has permission or a skill set, find a person who *can* update the firewall rule or clear the cache, bring them into the room, and have them pair up with a team member. The more actions you can have in your direct scope of capability in the crisis engineering center,[viii] the more likely they are to get done in the manner you expected. You can also then immediately surface any discrepancies or unexpected results back to the team.

TAKING ACTION DOES NOT MEAN TAKING EVERY ACTION

Beware of the temptation to resolve everything at once or to accept a convenient story that simplifies the problem to the point it becomes invisible. Instead, use the insights you've gained to guide deliberate, impact-making steps. The sensemaking process doesn't end here; it evolves as the system evolves. Each action you take is a step toward greater understanding and greater control—and that is how you create lasting improvement.

It is tempting, especially in the face of overwhelming dysfunction or crisis, to leap to dramatic conclusions and sweeping initiatives. And it's extremely common: Research shows that people fall back on "automatic responding" whenever they lack the ability to fully analyze a situation.[4]

But such moves—hiring a thousand new staff, commissioning a multi-million-dollar rewrite of legacy software, or launching an immediate organization-wide training overhaul—often bring more harm than good. Impulsive decisions can backfire; you need to strike the right balance between urgency and risk. Repeated, Goldilocks-size actions from the crisis engineering center are the better way to proceed.

While we encourage novel actions, we don't encourage just *any* novel action. The idea is to try something new that has some plausible way of making things better, even if trying it teaches you that

---

[viii] Anthony Downs observes that in a situation like the Cuban missile crisis, officials tend to advocate entrusting maximum responsibility to their own parts of the organization (*Inside Bureaucracy* [Little, Brown, 1967], 106).

it does not. Smashing the printer with a baseball bat à la *Office Space* may be novel, and may be cathartic, but it definitely won't help.

STAY OPEN TO NEW INFORMATION—AND PIVOT

Every action you take will reveal a new person, step, data point, exception path, or other information that you did not have before. You will be best served if you update your map accordingly, rather than try to debate the shift in circumstance.

If you want to be successful, you must be open to the idea that you are wrong, that your explanations have been wrong, and that your last decision or action was wrong. This way of thinking gives you the freedom to pivot quickly. But be warned: Our brains don't like to do this, as the desire for consistency is a central part of human behavior. Researchers have found again and again that we will act against our own best interest in order to feel and be seen as consistent with our prior statements and actions.[5]

Remind yourself and those around you to stay open to new information, and to update your maps, plans, and actions accordingly.

ANALYZE RISK AS YOU TAKE NOVEL ACTIONS

> On a big fire there is no time and no tree under whose shade the boss and the crew can sit and have a Platonic dialogue about a blowup. If Socrates had been foreman on the Mann Gulch fire, he and his crew would have been cremated while they were sitting there considering it.
>
> —**NORMAN MACLEAN**, *YOUNG MEN AND FIRE*

Crisis engineering demands a careful balance between rapid action and strategic thinking. You are always operating on multiple time scales at the same time, and one of those time scales is always "right now, today." While urgency is paramount, reckless decisions can exacerbate problems or create new, undetected ones. The purpose of sensemaking during a crisis is to act—and to act quickly. However, effective action requires clarity, focus, and prioritization. This section touches on how to navigate this balance and make informed decisions under pressure.

This may not be intuitive, but the *less* experienced at crisis engineering your organization is, the *less* time you should spend analyzing risk. An organization more experienced at crisis engineering can spend more time analyzing risk because it has (more) accurate maps of its systems and knows its stakeholders at least somewhat well. If your organization lacks these foundations, or is operating mostly from inaccurate maps and data, extensive analysis won't buy you very much, because you don't have any actual capacity to assess risk or model the outcomes of actions. In such cases, taking novel actions based on whatever available information you do have is the best course.

In the face of unknowns and ambiguity, if there isn't a clear best action, guess and check. You will always learn something that will contribute to the next guess and action. As Richard Cook says, "All practitioner actions are actually gambles."[6]

Types of Risk

There are various types of risk to consider during a crisis. Each type of risk has different implications and requires different mitigation strategies:

- **Political Risk**—What are the risks related to your political capital with your leadership? This is especially relevant in governmental and highly regulated industries, but political capital is relevant everywhere. Could the wrong choice lead to a loss in a governmental election, a board election, or a key leadership post?

- **Technical Risk**—System failures, data loss, or the unintended consequences of technical changes. Changes that remove resilience, such as eliminating redundancy or adding a dependency on a single-vendor product, are another important category of technical risk.
- **Reputational Risk**—This includes how actions or inactions affect public perception, customer trust, and brand value, as well as the perception (and even future) of the individual leaders involved.
- **Human Impact Risk**—Consider the psychological and emotional effects on employees and stakeholders. Too much change or poor communication can cause undue stress and trigger mass resignations, union negotiations, and/or lowered productivity.
- **Risk of Inaction**—Delaying action or not acting at all can both be seriously risky.
- **Financial Risk**—This includes costs due to fines, lawsuits, and/or lost business, as well as impacts on current and future budgets.

Identify Risk

Risk identification is an iterative and collaborative process. To uncover, and plan for, potential risks, consider these actions:

- **Simulate Risky Scenarios.** Bring together the actual decision-makers—literally around the same table if possible—to act out the intended action. What happens if it goes well? What might go awry? What are those mitigations? Tabletopping, where participants sit around a table and act out the steps of a hypothetical scenario one by one, is a skill that can serve your organization well the more you practice it. For example, walk

through what would happen in a hurricane, an extended power outage, etc.
- **Talk to People.** Consult key individuals impacted by the proposed action, especially the ones who would actually carry it out. For example, if your proposed action is to skip a processing step in the intake office, did you talk to someone in that office? Will they feel too nervous or risk-averse to actually carry out this change? Will skipping this step prevent them from accessing the next screen in their workflow? What mitigation could you put in place that would make them comfortable?

Reduce Risk

Mitigating risk is as important as identifying it. Here are some practical strategies:

- **Plan for Rollback.** Identify how to undo changes if they don't work out. Be liberal with reversible actions but cautious with permanent moves like hiring new staff or making public commitments.
- **Start Small but Strategically.** Take incremental steps that deliver quick wins and build momentum. For instance, instead of overhauling a legacy system, start by automating one repetitive task or resolving a specific bottleneck.
- **Accept Responsibility.** Shift the risk onto yourself or your team where possible. Taking ownership can reduce the blame on others if things go wrong.
- **Take One Action/Risk at a Time.** Focus your efforts on a few high-priority actions. When many actions happen at once, it can be challenging to parse which ones had which effects.
- **Communicate Clearly.** Transparency fosters trust.

Keep stakeholders informed about your actions and rationale to mitigate reputational risks.
- **Iterate and Adapt.** Risks are not static. What seems like a manageable issue today could escalate into a crisis tomorrow. Be prepared to refine your approach as new information emerges. Iteration is essential in crisis management—not everything will work on the first try.
- **Monitor Key Indicators.** Track relevant metrics to gauge the impact of your actions, such as call center volume or health metrics dashboards. A sudden spike or drop in part of your process might be a sign of success, or it might be a sign of an unintended consequence.
- **Use the Pareto Principle.**[7] Focus on the 20 percent of actions that will deliver 80 percent of the results. Prioritize taking one action that you expect to have the biggest impact on the problem, rather than five separate ones that collectively might have the same impact. Not only is it easier to take one action instead of five, but it will be clearer to see if the action actually fixed anything. If one of the five smaller actions fails, now you've got to troubleshoot across all of them.

Oopsies, I Made a New Problem

Unintended consequences happen. You're sensemaking! When they do, focus on damage control and responsiveness:

- Assess the impact of the new problem.
- Document and communicate about what went wrong—what assumptions need correcting?
- Take a new, corrective action based on what you learned.

At the end of the day, don't be paralyzed by fear of what might happen. Take action! It's the only way to learn the realities of your environment and to emerge from the crisis successfully—if not stronger. Even if the first action doesn't work out as expected, it's almost certainly better than inaction. Crises are dynamic, and progress comes from moving forward, learning, and adapting.

Next, we'll talk about how to measure how close you are to the end of the crisis—which will, in turn, help prioritize what actions to take.

CRISIS CHEAT SHEET

Crisis engineering is about creating conditions for effective sensemaking, and the beating heart of sensemaking is taking actions on a system and observing the results.

The crisis engineering team must repeatedly take novel actions. This is what leads to procession through the crisis, a more accurate understanding of the complex system, and more successful long-term outcomes.

Opening Moves

- Change one word on a screen of an affected system.
- Talk to operators who perform routine operations and ask about scheduled service disruptions, management behavior, gaps between procedure and practice, shortcuts, and fragile parts.

Take actions that are naturally easy and remember that humans provide the adaptive capacity in complex systems.

Candidate Novel Actions

- Turn it off.
- Try a process in a new way.

- Eliminate or change steps.
- Add or remove friction.
- Batch-process certain classes of items.
- Move people around in a process.
- Add automation.
- Rewrite instructions in plain language.
- Enable self-service.

Removing something is always a novel action.

Useful Tools for Taking Actions

- **Changing load on components:** load balancing/traffic routing systems, blocklists/allowlists, options for providing degraded operations, planned downtime
- **Modifying/hijacking component inputs and outputs:** pluggable pipelines, batch processing and CSV files
- **Modifying components:** fast/no-gates release processes, static content (instead of dynamic content), strangler pattern, turning off features
- **Quickly creating new components:** spreadsheets, scripting languages, scheduled jobs

If it is not immediately clear who should take a given action, then it should be a member of the crisis team. They should find the person with the necessary access and pair up with them.

Do not take every novel action. Dramatic conclusions and sweeping initiatives usually harm more than they help.
Stay open-minded to new information and pivot.

Analyze Risk

- **Types of risk:** political, technical, reputational, human impact, inaction risk, financial
- **Ways to identify risk:** simulate risky scenarios, talk to key individuals who are impacted by a proposed action
- **Ways to reduce risk:** plan for rollback of any action, start small and strategically, accept responsibility, take one action at a time, communicate actions and rationale clearly, iterate and adapt, monitor key indicators, use the Pareto principle

Some actions will have unintended consequences and that is okay. They still lead to learning.

2.6

Manage the Story

> A story that honors the dead realistically partly atones for their sufferings, and so instead of leaving us in moral bewilderment, adds dimensions to our acuteness in watching the universe's four elements at work—sky, earth, fire, and young men.
> —**NORMAN MACLEAN,** *YOUNG MEN AND FIRE*

We've taken a close look at the human and machine interactions in a couple of bona fide crises. We've stood up a crisis engineering center, we understand the basics of mapmaking, and we know who we're looking for. We have hopefully persuaded you that taking novel actions will get you further, faster, than lengthy deliberation.

Now, we'll weave these skills and techniques together in a way that consistently works to move the crisis toward resolution. This section can also help you decide what novel action to take first.

These techniques have one thing in common: They improve group sensemaking. Your goal is to assemble facts into a plausible story of what's happening in the crisis and construct a shared explanation for guiding further action. To determine which facts to use, you or someone else must first *notice* something.

Often, the thing you first noticed will become the end of the event you are constructing a story around. This is because events in a complex system tend to precipitate events with bigger and bigger consequences until one becomes noticeable.

While this one noticed fact is helpful, you don't know yet whether you are looking at the beginning, the end, or something in the middle. You will need multiple facts to construct your story. You will find these facts by taking actions based on your best guess of what is happening and what will make it better, and continue a guess-and-check loop of this behavior as your environment crystallizes into focus. This can go on forever. Every question that gets answered creates more questions.[i] Nonetheless, for humans to perceive a story, it has to have a beginning and an ending.

We will call the process of deliberately choosing beginnings and endings **bracketing**. It is one of the most powerful devices you have to accelerate sensemaking.

Noticing and bracketing can effectively direct the attention of a group without anyone ever appearing to be in charge. When people are operating on the same facts, it reduces conflict and wasted effort. Anyone in the room can suggest reasonable next steps. Actions happen without struggle. The people are even said to be on the same page.

Noticing and bracketing skills are worth developing. They work at any time scale. They can work regardless of your role. And they can be used outside of a crisis engineering effort.

---

i   One of the important points Robert Pirsig makes in his book *Zen and the Art of Motorcycle Maintenance* (Mariner Books, 1974).

> It may not be intuitive, but people are amazingly good at keeping track of nested stories. As long as the nesting is structurally sound (that is, each story is entirely contained in one other story), the sky is the limit. It is annoying and unsatisfying when the nesting is done wrong, even if we can't quite figure out why.
>
> An example is the 2003 movie *The Return of the King*. By the end of the film trilogy, many story lines are left open: The Ring must be destroyed, the war needs to be ended, Frodo needs to return home, and the other hobbits go back to their lives. The way they are resolved doesn't correspond to the way they were introduced. So the movie seems to "end" over and over.
>
> A counterexample came out in 2001: *Ocean's Eleven*. The first plot introduced is that Danny Ocean has just gotten out of prison and wants his ex-wife back. This leads to the heist scheme, which leads to recruiting challenges, then many, many substories within the heist. All of them are tidied up and resolved in the reverse order they are introduced, making a movie that viewers can follow and root for. (Even if it takes multiple viewings to really catch every detail.)

WHAT IS WORTH NOTICING?

When you have a complex system, a great many things are happening at once and are constantly changing. This makes noticing a surprisingly subtle skill.

Noticing, the act of separating a piece of information from the background, takes two essential judgments. The first is to say that this information is significant. **Significance**, for our purposes, means that we think it has something valuable to tell us. The second judgment assigns a value to the information: "Is this good or bad?"[ii]

Is It Significant?

Trained intuition will help you determine significance. If you are trying to perform a feat of crisis engineering, you must have some connection to the system in question. You might be the subject matter expert who has interacted with this production process for a long time. You might have a particular technical domain expertise, such as being the person who understands the new monitoring tool. Or, you might be the fancy crisis engineering consultant who was just brought in; you don't know anything in particular about this system, but you have seen others that operate according to the same laws. Any of these roles will provide you with a context that will help your automatic System 1 point out weird things.[iii]

If you truly have no background on the system in question, you may still be useful if your trained intuition of social dynamics is good enough. All groups of humans have blind spots, groupthink, and undervalued members. Merely being a neutral observer can add value.

Is It Good or Bad?

After you have identified something significant, your trained intuition has probably also given you a feeling about whether it is a good or bad thing. Be careful about assuming things you don't know at this stage. We would generally say something noncommittal, like

---

[ii] Other authors, including the ones we cite, use the words "scanning" and "framing" to mean "noticing," and "valance" to mean "value." We are using the words that are closest to plain English, even if it doesn't entirely agree with the jargon.

[iii] Even Daniel Kahneman's apparent nemesis Gary Klein agreed on this point, which led to their joint paper: Daniel Kahneman and Gary Klein, "Conditions for Intuitive Expertise," *American Psychologist* 64, no. 6 (2009): 515–526, http://dx.doi.org/10.1037/a0016755.

"This may be a clue to what's going on" or "This may become a problem." It may seem best to be totally neutral and describe the anomaly as "not good or bad, just something to think about." But something that is neither good nor bad tends to get ignored, because it doesn't motivate anybody to care about it. It is better to speculate and be ready to change your mind later. Anyway, let's be real—it's a rare day when an anomaly in a complex system is a good thing.[iv]

WHEN DOES IT BEGIN AND END?

Having generated interest in the thing you noticed, the hard part begins. If you're lucky, someone's natural curiosity will cause them to investigate. Curiosity, in these scenarios, will often be powered by a fear of being found at fault, which is ever-present and sometimes useful. If the mystery involves the database, for instance, and if you have a database vendor inside the crisis team, you will find them highly motivated to demonstrate that the problem is somewhere else, *whether or not this is true.*

This causes many anomalies to mysteriously disappear in the early stages of investigation. Problems are found and quietly fixed. We write it off and move on. Everyone can guess what happened anyway, and pushing a contractor (or anyone else) to admit a mistake is usually not worth it. If you are doing a halfway decent job, it's not the crisis team that they are afraid of anyway—it's their anxious bosses back at home. They'll probably tell you what really happened if you can talk to them in a situation where their bosses aren't around: in an elevator, in the bathroom, in line for lunch, or in the parking lot. (This is yet another reason not to allow an audience in the crisis engineering center *and* to walk around the floor as you map.)

If the anomaly does not go away quietly, you are going to find

---

iv "Complex systems contain changing mixtures of failures latent within them" (Richard I. Cook, "How Complex Systems Fail," https://how.complexsystems.fail/#4).

yourself reviewing a lot of evidence that suggests places the problem is *not*. This can be difficult. Complex systems always contain a mixture of ongoing failures and errors. Nearly every team approaches every interaction with the basic assumption that their area is good and right and true, and the other teams are hot, buttered garbage.

It will test your patience, but it is critical to be seen as neutral in negotiating cross-party debugging. The minute you are not, you become just another combatant, and progress will stop until someone else assumes the role of neutral arbiter.

Your job while guiding such an investigation is to maintain the story structure. You have found that anomaly X exists, and you are trying to find the answer to why X is true and what X means. There will be submysteries that need to be solved. But without constant reminders of the current goal, groups will wander in many directions at once. Ideally, you can have the question and subquestion (and sub-subquestion, and…) written on the whiteboard and visible to everyone in the room. If you can't do that, you have to find a substitute. We have been known to curate a Sharepoint folder full of documents with names like "Scooby Doo Mystery #7: The Case of the Disappearing Mail."[v]

Usually, it is good to follow one thread at a time, but there is an exception. In the early stages of working on a truly screwed-up system, it may be productive to let everyone look for targets of opportunity. This means letting them notice their own anomalies and pursue them in parallel. In this case, you need to rebracket the whole day's effort as "We are doing a general cleanup, and here is how we will know when we are done." Expect to introduce a new story structure before everyone is ready to stop exploring. They will not be ready at the same time, but it is impossible to hold the attention and urgency of people who are waiting around with nothing to do.

Part of maintaining the story structure is declaring endings at the appropriate times. Ideally, it is self-evident when you have found the answer to a question. Broadcast that answer so that everyone hears it (and write it down in the shared journal).

---

v   This is a real example!

Not all questions will end this way. You may hit a limit where you don't have the access or capability to take the next steps. The anomaly may disappear mid-investigation, leaving no trace. Or it may become clear that you are going to have to settle for a partial solution. You will need to perform the function of closing debate. When that happens, be ready to direct attention to the next story. A strong candidate is: "How are we going to work around the anomaly we can't address?"

If no one is tending to framing, naming, and bracketing, troubleshooting efforts degenerate into random spitballing and boredom. You will not be able to hold people for more than a few days if there is no story progress. They will report to their bosses that they are wasting their time here, and they will be correct.

GET UNSTUCK

Another duty you have as the story maintainer is to make sure the action keeps moving. We have all read books with pacing problems. *Dune*, by Frank Herbert, a story that is generally regarded as successful in other ways, has Paul racing off into the desert to hide with little else happening for a very large number of pages. More than one reader has gotten bogged down and never finished.

There will be times when an experiment has failed, which means it yielded a result that added nothing to your understanding, and there is no obvious next step. There may be other hypotheses or experiments on the whiteboard already that you can jump to. You may have to supply a clever idea of your own, if the problem is familiar. If all else fails, there are some general techniques that have been proposed in psychology textbooks since at least 1974:[1]

- **Reverse "Good" and "Bad."** Usually, "good" and "bad" are weakly defined on the inside of a complex system anyway. Is it good that a particular claims-processing office has a backlog that is three times longer than its peers? Seems like no, but there may be an explanation we are missing. We might find that explanation if we reverse polarity and look for "Why are the other offices' backlogs so short?"

- **Relabel Elements with New Meanings.** Names and boundaries inside a complex system also tend to be malleable. If you have spent an unreasonable amount of time disagreeing over whether an anomaly originates "in the database," start asking exactly what each team means by "in the database." The answers may reveal a gap where the problem is hiding. If nomenclature needs to change, it is always easier to make up a new name and insist that people use it than to get either side to back down. You are allowed to make the name comically verbose if you like. We have all diagnosed problems in The Base System That Stores Data But Is Not a Database.
- **Try to Accomplish the Opposite of What You Want.** Similar to reversing "good" and "bad." Oftentimes after a protracted struggle to get a piece of code to do what we want, we will try to see if we can make it do something even more wrong. The point is to see if anything we are doing is affecting the behavior at all. "Try turning it off" is another such test. When the answer is "No, you cannot," you have learned big facts.
- **Disregard Factors You Cannot Change.** You want this one to be obvious, but sometimes you will have to fight for it. Layer Aleph did two large projects during the COVID pandemic. One was working on unemployment-claims processing, and the other was related to COVID vaccine data collection. In both cases, we found leadership offices paralyzed by pointless debates about what was going to happen to next week's volume of benefits applications or vaccine doses. No one knew any answers, let alone had any control. We had to argue many times to

redefine goals into things that were good to do no matter what happened to the counts.

Equally important is the other end of the bracketing exercise. Don't wait for everything to get back to normal, which will never happen. If it's possible for things to go back to normal, you didn't have a proper crisis to begin with. The effort to deal with the crisis, however, will definitely have an end.

Because things don't go back to normal, people have a hard time recognizing when the crisis has ended. Our advice on how to wind down a crisis engineering project is to consolidate what gains you can and reset expectations so as to be ready before the next crisis. Projects that are not precisely "crisis engineering" need closing rituals as well. It's nice if it can be symmetric with the opening ritual, but that won't always make sense. There were several hundred people at the outset of the HealthCare.gov project, but the core team that was left in the XOC by the end of the enrollment push were a couple dozen—small enough to fit in the DoubleTree hotel bar.

The crisis won't be remembered as a success if it doesn't have a story, and it's not a story without an ending.

THE PRICE OF POOR BRACKETING IN COVID-19

Three Mile Island helped us understand sensemaking by considering a situation where it was absent. Similarly, the COVID-19 pandemic makes for a great demonstration of how important bracketing is, because it had none.

The pandemic began in…when was it, exactly? Most Americans became vaguely aware of this highly contagious disease by March 2020. There was no one news story, or one date. By the time DC health authorities acknowledged "community spread," it had been happening for weeks.

In the early weeks that could have sustained a call to action, we had President Donald Trump repeating that it's "just going to disappear."[2] Meanwhile, public health officials admonished Americans to *not* wear a mask, which was approximately the only action they could possibly take. As the months dragged on, public health orders came and went in different places at different times, with no consistent rationale. Most of them were a call to further inaction: business closings, limits on gatherings, and stay-home orders. Most people had a part of their identity and role taken away from them, with nothing to take its place. This provoked severe backlash, and a feeling among many people that the pandemic was being done "to" them by authority figures. The damage to the social fabric is still unspooling today.[vi]

When did the pandemic end? A few readers are now screaming into the page that it's still going on.[vii] For around half of Americans, it ended when they got the vaccine, which was a messaging strategy by the Biden administration that labeled the subsequent waves a "pandemic of the unvaccinated." For most of the remainder, the vaccine was too late to play a role in the story: Their pandemic had ended long ago, when their patience for "lockdowns" expired. This is how we came to see large numbers of Trump supporters reject the vaccine that the Trump administration developed.

This was a story that had no beginning and no ending, and no consensus reality will ever exist.[viii] The number of American lives lost in the first few years (over a million, in CDC estimates) approximately equals the number lost in all wars combined. Few people rate the response as successful on any metric. The day before this chapter was first written, President Biden issued a preemptive pardon to his former chief medical advisor, to forestall future prosecution. These are the consequences of letting national sensemaking fail so completely.

---

vi At this writing, vaccine resistance that spread during COVID has put measles back on the map.

vii We are confident this is true no matter what year you are reading this book.

viii For example, one way we could have emerged is by centering the knowledge that by addressing air cleanliness in buildings, we could virtually eliminate airborne respiratory diseases like COVID the same way we did waterborne diseases like cholera.

CRISIS CHEAT SHEET

To maintain good storytelling as an incident lead, pay attention to:

- **Noticing:** Separate an interesting fact from the background and call attention to it.
- **Bracketing:** Deliberately create groups of related events with beginnings and endings.
- **Naming:** Make it possible to talk accurately about facts, events, and groupings by giving them unique names.
- Look for interesting facts worth noticing.
- Begin new stories with a clear question or problem statement.
- Do not let action get stalled. Try the following if stuck:

  1. If you don't know what causes an anomaly, ask why it usually *doesn't* happen.
  2. Reexamine system labels; see if everyone is using them to mean the same thing.
  3. Test whether you can make a problem *worse*, to find out if your actions are having any effect at all.
  4. Rule out environmental factors outside your control.

- Clearly communicate when a question is answered or a mystery is solved, and follow up immediately with the next question.

2.7

Measure Progress

> So it came finally not in God's time but in the considerably slower time of bureaucracies, yet it came.
> —**NORMAN MACLEAN,** *YOUNG MEN AND FIRE*

As you're taking novel actions and updating your map to better reflect reality, you also need to measure the impact those actions are having (or not) on ending the crisis. Increasing uptime or timeliness may not have any impact on a crisis caused by a high error rate, just as clearing a report backlog may not help production get new products out the door any faster.

Measuring progress in a crisis isn't just about gathering data—it's about using it to create a shared story with your entire team, against which you can take new actions to improve the state of affairs. This is *the* shared story you'll ultimately want to spread across the entire organization—and the one you can use in a crisis to fundamentally shift the entire organization's priorities going forward.

In a crisis, these measurements need to go far beyond knowing if a system is simply operational. Success can be nuanced: It's not only about whether things are working for its users now but also about predicting stability in the near future. Sure, it's up now, but how confident are you that it will be up in five minutes? If the backlog shrinks today, is that due to effective interventions or simply lower weekend traffic?

Progress measurement can help answer these questions and confirm whether your novel actions are yielding the desired outcomes.

SETTING HEALTH MEASURES

We've almost never been on a gig where the organization had an objectively measurable goal for emerging from the crisis. What measures indicate you're in a crisis right now? What needs to change for the crisis to be over and the organization considered healthy once again?

Define Your Goal

What does "up" mean for your system? What amount of slowness or downtime is tolerable versus crisis-level? A few other questions to consider:

- What does "operational" mean in this context?
- What metrics will indicate you've achieved your goal?
- Who calculates these metrics, how often, and with what methods?
- What time horizon do the metrics cover (e.g., last hour, last year)? The longer the horizon, the harder it will be to impact quickly.

For instance, there is a chain of fast-food restaurants called Waffle House. The unofficial "Waffle House Index" gauges the severity of a disaster by the operating status of local Waffle House restaurants:

GREEN: Restaurant is open, and full menu is available.
YELLOW: Restaurant is open, but menu is limited; this may be because of limited supplies and/or electricity.
RED: Restaurant is closed.[1]

This is clear and elegant—so much so that FEMA uses it in their disaster response planning.

When looking at progress through the lens of this example, we'd focus on:

- **Backlogs**—Are we producing waffles faster than they are being consumed?
- **Process Times**—Are people getting their waffles in a reasonable amount of time?
- **System Metrics**—Are the waffle irons operating within acceptable temperature ranges?
- **Exception Handling**—Are kitchen spills and fires being handled properly and in a timely fashion?

Limit measurements to those that align with your crisis goals. Too much data or too many goals will overwhelm and distract the team. It's fine to rely on proxies for the overall health of the complex system at first, like total claims processed per hour, orders cleared per hour, or total error counts at some important input or output node to the system, and so on.

Defining and measuring progress between the current and goal states is crucial, because it gets everyone looking at and working toward the same story—sensemaking in action.

This may require creating new metrics. Perhaps you've discovered a signal that would have alerted the crisis earlier, had anyone been measuring it before. That's a good starting point.

Be realistic about the goals you set. A crisis presents a rare opportunity to fundamentally shift one or more measures that the organization cares about. Don't set yourself up for failure by setting a metric like a backlog of zero, when a 5 percent backlog may be a more realistic and manageable number, but still transformational relative to the size of the pre-crisis backlog.

> Watch out for Gresham's Law:[2] Work that produces measurable outcomes tends to drive out work that produces unmeasurable outcomes. We cringe when we hear of technical organizations counting lines of code written or number of commits as a proxy for technical skill, which results only in driving out your best technical talent. Pause to consider whether goals you set are the *right* ones or they're simply the easiest ones to measure.

SENSEMAKING WITH DATA

Is it possible to have accurate, comprehensive data and still have a crisis? Sure. But we generally find missing, inaccurate, or undefined (interpreted differently by different teams) data to be a culprit.

Start with the raw data you have about where you are relative to your goal:

1. **Run the Data.** Look for anomalies and inconsistencies.
2. **Audit Calculations.** Verify formulas and assumptions with your own eyes. Look at code and formulas, not manuals or documentation.
3. **Prioritize Data People Know Is Missing.** Examples include missing monitoring for website errors, like in the early days of HealthCare.gov. If you're missing this level of instrumentation, you need to get it ASAP. Prioritize the procurement and IT approval for the tools, dedicate budget, and make space and time to set it up immediately.
4. **Cross-Reference Results.** Validate outcomes by taking novel actions to confirm (or disprove) them.

What if you can't count it? Get creative about your measurements. We've measured backlogs in inches of dot matrix printouts. If you can't count records, how else can you know if it's growing, shrinking, or staying the same? Avoiding counting altogether prevents identifying some of the most problematic areas of the process.

Remember, it's practically guaranteed that any initial data you're working with is flawed in some (or every!) way. To keep improving your shared understanding of reality:

- Trace data from its original, unprocessed source and follow its path in granular detail to identify any unwanted or unknown transformations. Understand every manipulation along the way. Where does the data come from—is it imported or hand-typed? Is there code (e.g., stored procedures, macros) applying transformations?
    - It's rare that one person understands this entire data flow (prior to a crisis, anyway), so it's important to look at the actual computer screens with the people whose hands touch that keyboard.
    - Be on the lookout for data points that are not yet being considered. Examples include hidden wait lists or uncounted mail. You can usually identify this when you stumble upon a storage room as you map or as you compare your map to available data points.
- Understand who owns, who can pull, and who can manipulate the data. Operators might call manipulation "processing," "cleaning," or "verifying."
- Hold off on manipulating data by merging, totaling, counting, or doing anything else that implies analysis you didn't actually confirm with your own eyes, until you have raw data that you fully understand.
- As you make changes, cross-reference data against process maps and system diagrams. Are counts consistent? Are you missing data for certain systems or steps? Do multiple steps/systems share definitions of shared concepts (like a case, user, or transaction)?

- Run counts by your new friends in the field. Do they seem right? If not, do these people have an intuition as to what's off?
- How does this analysis look against different time horizons? A change you made an hour ago may be making huge improvements that you won't be able to see on a weekly or an annual report.
- Identify lags when interpreting trends. Different data sources operate at different speeds. Technical systems generally have real-time monitoring, while customer support or finance may have infrequent, periodic updates. Don't compare the two if the former updated one minute ago and the latter won't update for three more months.
- Is a system consistently down or slow, or is this an unusual issue?
- Are surges due to operational problems or predictable events (e.g., Black Friday, 2 a.m. database backup, February has fewer days than other months)?
- Is the error code actually an error or an informational alert?
- Are duplicates or outdated records inflating or otherwise skewing metrics?

As you take action, regularly compare expectations against outcomes:

- What did you expect to happen?
- What actually happened?

Keep updating your map, journal, and calculations with what you learn.

As corrections cascade through datasets, expect resistance and/or confusion. A correction can cause many downstream numbers to change significantly, causing people to question *everything*—including many things unrelated to the crisis. Someone who spent the last seven years printing out a report showing a "green" count is not going to like seeing that stat turn "red." Leadership may balk at the true number of backlogged claims or exceptions requiring human handling, and try to persuade you to minimize it. This is a necessary part of the crisis engineering process.

As you continue refining measures, be prepared to rebuild confidence in the new numbers while being open to the idea that your new numbers could *also* be wrong.

Automate Measurements

Measuring progress toward your goal will be highly manual at first. After all, you don't yet know if you're adding it up correctly, missing giant puzzle pieces, and/or double-counting items. This is approximately the last place you want an AI to "hallucinate." You have to do this manually, because that's how sensemaking happens.

But as you progress, you can start to automate repeatable measurement processes to maintain consistency, reduce manual errors, and prevent similar crises in the future. This can start as a spreadsheet or command line script.

To get automation right, we've had clients spend *months* after the crisis running them over and over, identifying issues, resolving those issues, trying again, and fixing the next issue, in order to land on a truly reliable process. This is normal.

Don't worry about decimal-level precision or how many seconds "real time" means unless your operations demand it. A daily pull of data is generally enough for operational information, and seconds are probably fine for technical systems.

Automation will generally progress through these phases as your crisis response evolves:

1. **Basic Monitoring**—You know for sure if the system is on or off.

2. **Ad Hoc Reporting**—Help desk representatives can report anomalies to a team that can investigate them.
3. **Standardized Metrics**—Reports are automated and objective, not handcrafted or easily manipulated.
4. **Advanced Insights**—You can track trends based on reliable, widely understood data with no gaps.

You can probably emerge from a crisis while still in the first phase of measurement, but a real crisis engineer has their eye on making stage four the new operational reality.

THE HUMAN SIDE OF DATA

Data usually has a human element, including how it's reported, input into the system, pulled out of the system, and interpreted. Considering this human element can prevent you from relying on data you shouldn't, and can help you design better ways to measure reality.

Perverse Incentives

Crises often expose organizational incentives that encourage hiding or distorting information. A Soviet factory, measured solely on production quantity, made exclusively left-footed shoes.[3] A bounty placed on the heads of dead cobras, intended to cull the population, instead encouraged people to *breed* cobras to earn money.[4] Measuring postal workers on timeliness led to the intentional disposal of late mail, so it never arrived at all.[5]

To counteract perverse incentives, think like your evil twin: How could you manipulate or achieve this measure in a way that's counter to the spirit of the goal? Develop countermeasures to catch and discourage this manipulation.

CONTROL VIA OBSERVATION

It is annoying but true that all efforts to observe a system alter its behavior. In this way, an electron and the U.S. Department of Veterans Affairs are the same. You can't change this fact, so use it to your advantage.

Suppose you are minding your own business on your remote ranch in Arizona and a white truck with a government license plate pulls up at the door. The person who gets out asks you the following series of questions:

- How much of your well water do you drink?
- Have you had the well tested for arsenic?
- Have you been experiencing any headaches or skin lesions?
- What would you do if the water were contaminated?
- How much can you afford to spend on a water treatment system?

Then she thanks you for your time, jumps back in the truck, and races off in a cloud of dust.

What do you do now? Nobody told you to do anything; they were just asking questions! You could go about your business and forget about it, but you probably won't. Most of us would see about testing for arsenic, at least.[i]

Humans provide adaptive capacity in part because they are great noticers. Evolution has trained us that when something unusual happens, it's good to find out why. Being asked new questions by an authority figure is an unusual event. Our reflex will be to protect the status quo by making the answers be uninteresting (or what we think is uninteresting, anyway). Without having ever heard of Mr. Beer or Ms. Leveson, we instinctively realize that the new lines of

---

[i] This is an exaggerated story of a real data collection done by the University of Arizona. Mikey's well water contains eleven micrograms per liter. This is fine, everything is fine.

questioning are the sensors of a control loop that is about to complicate our lives.

History has many similar observations. A British economist, Charles Goodhart, wrote a paper in 1975 that is usually paraphrased as Goodhart's Law: "Any measure that becomes a target ceases to be a good measure." Anthony Downs wrote about the inherent distortions in data collection through the control hierarchy. We would make a stronger statement than either of them, which is that all data collections that are known to humans will be manipulated.

There are upsides. Making a problem visible sometimes causes it to be solved with no further action from you. This even works *better* when you have weak accountability and organizational rivalries. If you have a dozen vendors that are all entrenched in their belief that their part of the system is working perfectly, and you create a dashboard that says differently, and make it visible to all, the cognitive dissonance will be intolerable. Even as they complain that the data isn't accurate, they will make the problems disappear.

WHO CAN SEE PROGRESS?

While at first, you'll want to keep numbers close to the vest—as you don't yet understand or trust the data—over time you'll want to use your success measures to create a widely shared story of reality.

Public dashboards can help disseminate key information to more people, including members of the public. The data may not be pretty, but remote staff knowing the system is down (and it's not their Wi-Fi or computer) can help eliminate a lot of back-channel gossip and confusion. It can help the workforce to reengage if they have a place to watch things improve. It will help everyone converge on a shared story.

You may also want to create specific dashboards for specific audiences, such as nontechnical leadership, so they can see key metrics (the system is working as well or better than expected) versus detailed real-time charts that can cause unnecessary alarm (for example, by revealing a known, nonproblematic spike at 3 a.m. caused by a scheduled database task).

Measuring progress is how you'll know how far you are from the finish line, and which actions are actually getting you there. Start with raw data, automate where possible, and remain vigilant against inaccuracies and perverse incentives. Soon, this will be a measure you can use to tell the entire organization a clear story about where you are and what you can achieve. Above all, focus on clarity and actionable insights to guide your crisis response effectively.

CRISIS CHEAT SHEET

You need to measure top-level information in order to judge procession through a crisis.

Set Health Measures

- Define a goal, considering what "up" or "operational" mean for your complex system, and over what timeframe (hours or days are more measurable than years).
- When looking at progress, focus on backlogs, process times, system metrics, and exception handling.

Sensemake with Data

- Run the data: Look for anomalies and inconsistencies.
- Audit calculations: Verify formulas and assumptions directly.
- Prioritize data people know is missing; procure and/or set up infrastructure immediately if you cannot measure basics like error rate.
- Cross-reference results: Validate outcomes with independent checks.

Deal with Noisy Data

- Trace data from its original unprocessed source to understand how it is manipulated.
- Understand who owns, can pull, or manipulate the data. Manipulation may be called "processing," "cleaning," or "verifying."
- Delay merging, totaling, counting, or other analysis until you have raw data you understand.
- Cross-reference data across your map. Look for inconsistencies and investigate them.
- Check counts with your new friends in the field and see if they seem right.
- Consider analysis across different time horizons.
- Identify lags when interpreting trends. Different data sources have different lags.

Automate measurements, progressing over time from basic monitoring, to ad hoc reporting, to standardized metrics, and eventually advanced insights.

Mind the Human Side of Data

- Prevent perverse incentives by thinking about how *you* might manipulate or exploit the current goal and incentive structure. Look for places where parts of the system may be exploiting them already.
- Remember that observation (gathering data) changes people's behavior. Exploit this if useful.

When you have a working understanding of your measurements of progress, share it in order to create a widely held story of reality.

2.8
Communicate in a Crisis

> One of the chief privileges of man
> is to speak up for the universe.
> —**NORMAN MACLEAN,** *YOUNG MEN AND FIRE*

Effective communication is a key part of your crisis engineering effort, but it is also a balancing act. On the one hand, you must ensure that the right people get the right information at the right time. On the other hand, you must avoid overwhelming your audience and prevent unnecessary panic. You need to communicate *while* assembling a team, constructing an initially unsteady map of the terrain, taking novel actions that may have surprising results, and setting and computing measures. This is a hard juggling act!

This chapter explores the goals, challenges, and methods of crisis communications, offering strategies for clear and actionable messaging with all your stakeholders, from leadership, to the members of the crisis engineering team, to users impacted by an outage or a delay.

A crisis communications toolkit can be built over time, so you have templates, messaging channels, status pages, and rules of thumb that you can deploy more quickly next time.

This chapter is not about crisis public relations—for that, we refer you to the inimitable Peter Sandman.[1]

GOALS OF CRISIS COMMUNICATIONS

Your goals of communicating in a crisis should be to:

- **Avoid Surprises.** Communications about current state, planned actions, and potential risks should prevent key stakeholders from being blindsided. For example, you don't want an executive learning about an outage via a call from a journalist.
- **Mobilize the Right People to Act.** Provide enough accurate details for those who have the ability to resolve the crisis, mitigate its effects, or provide critical information to take action. Not all helpful actions will take place inside the crisis engineering center, especially in larger organizations and/or at the start of the crisis response.
- **Avoid Unwanted or Unhelpful Actions.** Senior leaders who feel anxious or in the dark are more likely to make hasty, unilateral decisions like issuing a press release, firing someone, or hiring a swath of new people that will only add to your burden.
- **Avoid Alert Fatigue.** Alert fatigue directly caused multiple crises we've been brought on to solve. It can cause everything from patient death to accidental deletion of entire systems. Too many alerts overwhelm your team (and anyone downstream of them). They will start to ignore your messages, while potentially spinning up *their own* narratives.

CHARACTERISTICS OF EFFECTIVE CRISIS COMMUNICATIONS

- **Brief.** People do not read. We've found three sentences to be most effective. Three paragraphs made up of three sentences each is a good rule of thumb for a maximum length.
- **Focused.** As described above, different audiences need different information. Attempting to mash them all together into one update frustrates everyone. It's okay to say upfront what your communication will *not* include.
- **Timely.** There is a tightrope between alert fatigue and silence, while giving observers a predictable cadence of updates. Pull notifications (e.g., having someone go to a status page or chat room to check status whenever they think to, instead of sending hourly push alerts) can help with this, as long as the source is updated in a timely manner.
  - A clue that you're not updating people frequently enough, or predictably enough, is fielding many incoming requests for updates. Including the timing and/or circumstances that will dictate the next release of information in each communication helps reduce inbound requests and interim anxiety or speculation.
- **Prioritizes Pull over Push Alerts** (e.g., a status page people can check on their own as often, or infrequently, as they like), and ways to focus notifications only on actionable items. Look for, and suppress, noise from noncritical issues.
- **Leverages Existing Channels.** How are people currently receiving or submitting information about the crisis? Can you use this forum to provide updates that have low impact on staff or systems, such as adding a recorded message to the support call center, or a banner

update to the intranet? The crisis engineering center channel we discussed earlier is the best place to point people who have information.

Remember:

- **Silence Is Communication!** One of the worst things to do when something very bad happens is go silent. In the absence of communication from you, people will communicate on their own.[i]
- **Don't Communicate Using Infrastructure That's in Crisis.** If your website or network is down, the status update page on your website will also be down.

TELL THE TRUTH

When problems develop, some people's first instinct is to lie and cover them up. If you are one of them, you must suppress this impulse in a crisis. Your job is to facilitate group sensemaking, and spreading false information will do the opposite.[ii] Far more common is the behavior where *committees* of people convince themselves to lie and cover up. The danger is greatest when decision-makers are safely surrounded by layers of people that are certain they are the good guys and certain that they know better than anyone else. This structure describes most government agencies.

---

i "If a bureau is operating under great time-pressure, it will tend to use subformal channels and messages extensively, since there is often no time to check formal procedures and follow them. Thus, in a crisis, top-level decisionmakers will reach out for information whenever they can get it, whatever the channel structure involved. They will also tend to rely on other officials in whom they have great confidence, even if those other officials are not formally connected with the subject of the crisis (for example, Robert Kennedy's role in the Cuban missile crisis)." (Anthony Downs, *Inside Bureaucracy* [Little, Brown, 1967], 114).

ii A therapist may be able to help you understand what toxic environments caused you to develop this adaptation, but it's out of scope for us.

As you may have guessed, we have strong feelings on this topic. We can't say as much as we would like, because of nondisclosure agreements and also the fact that we won't have any clients if we develop a reputation for snitching. So we'd like to point you to the experiences of Dr. Peter Sandman. He is a retired professor of communications who made a career of managing crises, hazards, and outrage (roughly defined as the public reaction to a hazard, which may or may not be proportional to the hazard). He was part of the Kemeny Commission that followed up on the Three Mile Island accident, and worked with the federal government starting with the post-9/11 anthrax scare.

Dr. Sandman's lecture to the National Public Health Information Coalition in 2009 is in part a catalog of self-harms that the "good guys" have perpetuated in the belief that their policy aims were more important than telling the truth.

To this, he later added a new catalog of false public statements that were made during the COVID-19 pandemic:[2]

- Ordinary people have no need for masks, in March 2020
- Vaccinated people don't need masks, in May 2021
- Vaccine boosters are unnecessary, in fall 2021
- Unvaccinated people are causing most Omicron spread, in December 2021
- Natural immunity postinfection is not as good as post-vaccine immunity

We hasten to point out that this is not a list of statements that later turned out to be wrong given new information—there are many more of those, and public health can't be faulted for not knowing everything in advance. This is bad advice that was known to be either wrong or unsupported *at the time it was given.*

You may notice that most of the false statements were slanted to influence people to get a vaccine. The expert community believes in vaccines as the best tool for protecting public health, and for decades, it has been choosing messages to support that cause, rather than strictly following current facts. They have been caught countless times. As we write this, the backlash has accumulated into such a tsunami that vaccine denialists are in charge of a number of state health agencies, to say nothing of the U.S. Department of Health and Human Services. People who identify as Donald Trump voters are significantly less likely to accept the vaccine that the Donald Trump administration *developed*.[3] It's hard to overstate the damage that disingenuous messages have done to the cause of public health.

As Dr. Sandman puts it: "When you are trying to reassure people rather than alert them, then even small exaggerations—pretending to be 100% right when you're only 98% right—can do you in."

If you are in a position to affect the messaging around your crisis, avoid broadcasting those 98 percent truths as if they were 100 percent. It is much harder to help people when they don't trust you.

In many crises, there will be a strong temptation to take the easy road by putting out information that isn't quite accurate or complete. This is an illusion. It is never the easy road. Instead, try to remember the following:

- Don't over-reassure.
- Acknowledge uncertainty.
- Acknowledge disagreements.
- Don't aim for zero fear.
- Offer people things to do.

For further reading, these recommendations and many others are discussed on Dr. Sandman's website.[4]

CONSIDER YOUR AUDIENCE

We've seen both ends of the crisis communications spectrum. We've seen risk-averse teams say nothing, not realizing their silence is fueling media speculation and thousands of employee group chats. We've also seen technical teams send hourly detailed technical dashboards straight to the CEO, who, as a result of not being able to interpret the charts, and unwilling to ask anyone to explain it to them, takes undesirable actions to quell their resultant anxiety.

It may seem easier to take a middle ground and give everyone a high-level summary of information, but that doesn't work, either. You need a comprehensible, nontechnical explanation for nontechnical stakeholders, and you need enough details at high enough frequency for the technical experts to act on and respond to new information. In either case, you want to avoid overloading any one individual past their saturation point of being able to comprehend information.[5]

Audiences to consider different messaging for include:

- Crisis engineering team
- IT team responsible for affected service(s)
- IT team(s) (potentially) impacted
- Leadership
- Program staff
- Non-IT employees
- Impacted users
- Members of the public at large

Channels to consider using include:

- Phone call
- Text message
- Direct message
- Chat/channel
- Conference bridge
- Dashboard
- Status page
- Website banners
- Press release
- Mailed letter
- Emails to users

When developing your communications plan and cadence, ask yourself:

- Who needs to know now?
- Who needs to know later?
- Who will let them know? How will this person let them know?
- What will they let them know?
- What follow-ups are needed?
- What's being communicated anyway?
- Where are people informally communicating already?[iii]

As you communicate unpleasant information, consider the following:

- What do you need to say?
- Who needs to hear?
- Do they really?
- What are you afraid of?
- Who will argue against you?
- What could mitigate those arguments?
- Is there time to wait?

Technical leadership needs to take special care when communicating with nontechnical colleagues. Too many details risks boring or intimidating them, which will create a new cascade of problems to solve. Too few details risks making them anxious or doubtful about the veracity of the responsible teams' understanding of the situation.

As an example, can you tell the difference between these two images?

[iii] "The higher the degree of uncertainty inherent in a bureau's function, the greater will be its proliferation of subformal channels and messages" (Downs, *Inside Bureaucracy*, 114).

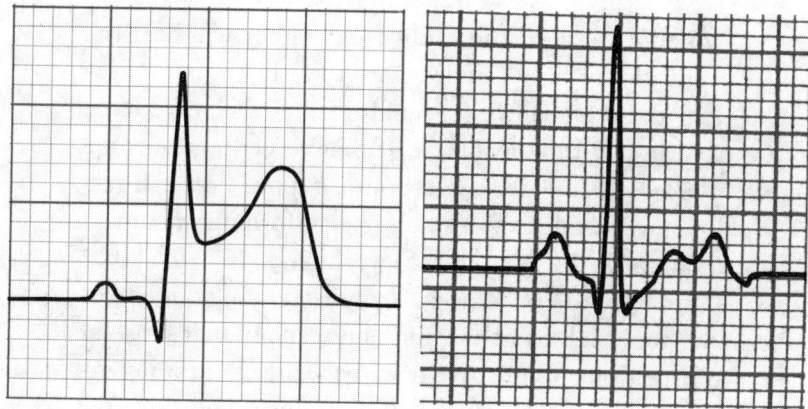

The left image is the EKG of a healthy heart. The right is an EKG of a heart attack. While a doctor may know to run down the hall at the second scan, someone without medical expertise (including the three of us) would be hard pressed to tell the difference.

This is akin to sending a network traffic diagram to the governor's office or the output of `strace -f` to the head of accounting.

Nontechnical explanations are a careful balancing act. There are few substitutes for practice. Notice which technical staff are more capable of delivering plain-language explanations that instill confidence. This will be the same staff who are more capable of collaborating outside the boundaries of the technical teams in the organization.

REFINE YOUR COMMUNICATIONS

Continue improving upon your communications throughout the crisis. Some strategies include:

- **Consider the Questions You Get After an Update.** Explicitly ask people receiving your communications for feedback and questions. Are people asking for the names of impacted servers? Include that information next time.

- **Collaborate with the Help Desk.** They will bear the brunt of any over-, under-, or confusing communication, and their call volume and reason logs are a source of near-real-time feedback. If they're getting a deluge of calls with the same question or report, how can you direct those callers to a status page informing them you know about the issue and are working on it?
- **Track Open/Visit Rates.** If you're sending out emails and they are mostly going unopened, that's useful information. If you have a status webpage, traffic patterns to it will give you a sense of how others perceive the urgency of the situation. This is infrastructure you can build and shore up for the future.
- **Pivot Your Tools.** Consider the communication channels you have available already. We don't suggest opening up a call center overnight, but you could probably open a company-wide chat, status page, or conference bridge quickly.

WIND DOWN COMMUNICATIONS

Any update about the crisis becomes noise after a while. (More on when to start winding down your crisis communications in chapter 3.2, Spin Down the Crisis.) Proactively spinning down your crisis communications ensures the right people continue to receive necessary information, without causing anyone to prematurely tune out.

- **Track Communication Pathways and Cadences.** Write down how and where information is being shared during the crisis, so you can leverage or thoughtfully redirect these channels going forward.
- **Ruthlessly Eliminate Updates.** If you're managing

your story, you have established tension. Now, as the situation stabilizes, release it as you reduce the frequency of updates. A common mistake is to establish very frequent communications during times of peak crisis circumstances, and allow them to persist for weeks or worse afterward. This mistake reduces the credibility of leadership and the crisis engineering team.[iv]

- **Put Up a Status Page**. Train users to look at the status page first. As long as it remains accurate and timely, this will significantly cut down on support calls and back-channel floundering as people wonder if it's the website or their own internet that's down.
- **Make Sure There Are Guardrails Around Future Use of Crisis Communications Tools.** If you set up a new company-wide alert email or a county-wide text notification in the middle of the crisis, make sure they're appropriately locked down so they aren't triggered in a test, or that any security controls missed in the flurry of the crisis are put in place to lock down access. At the same time, you don't want them so inaccessible that they can't be used in a future emergency. Regular practice exercises are the only credible way to strike this balance.

[iv] "Communication requires definite costs. Every message involves the expenditure of time to decide what to send, time to compose the message, the resource-cost of transmitting the message (which may consist of time, money, or both), and time spent in receiving the message. Also, if the message passes over a channel operating near its capacity, it may cancel or delay other messages" (Downs, *Inside Bureaucracy*, 112).

CRISIS CHEAT SHEET

The goals of communicating about your crisis include:

- Avoid surprises.
- Mobilize the right people to act.
- Avoid unwanted or unhelpful action.
- Avoid alert fatigue.

Characteristics of effective communication:

- Brief
- Focused
- Timely
- Prioritizes pull over push notifications
- Leverages existing channels

- Remember that silence is communication.
- Don't use infrastructure that's in crisis to communicate.
- Tell the truth and do not mislead to manipulate behavior.
- Acknowledge uncertainty.
- Acknowledge disagreements.
- Do not aim for zero fear.
- Offer people things to do.

Carefully consider any mismatch between the level of technical detail provided and an audience's ability to understand it. Refine your communications as you go:

- Consider questions received in response to outgoing communications.
- Collaborate with any local help desk.

- Track open/visit rates on any communications.
- Pivot your tools.

Don't forget to wind down your communications as the crisis ends:

- Track communication pathways and cadences.
- Ruthlessly eliminate communication updates as the crisis calms.
- Direct people to a status page if possible.
- Establish guardrails around future use of any new crisis communications tools.

2.9

When to Give Up

> Grant me the sense of proportion to judge the difference between an incident and a crisis.
>
> —AL-ANON

You've tried to get a senior official to declare a crisis, but they haven't. You've identified the team members and a room for the crisis engineering center. You've dutifully walked to every corner of the process and talked to every operator you could find to start a system map. Perhaps you've uncovered some concerning data and process breakdowns, and surfaced previously unknown insights from the field. You have an idea of what novel actions you'd take…if only someone would let you.

Despite the steps you've taken, it seems like everyone else is operating as business as usual.

This will be a very frustrating thing to hear—but it's possible that your organization is not in crisis at all.

Recall that we defined crisis circumstances as:

1. Fundamental surprise
2. Failure of sensemaking—perceptions break down, existing maps and models don't work
3. Degradation, disruption, or complete change of core processes or outcomes

4. High visibility—either internal to an organization, external to it, or both
5. Rigid deadline or timeframe

If you're feeling like Chicken Little, trying to persuade everyone around you that a crisis is happening while they're going about like nothing's wrong, review this list. Are you *actually* in a crisis? If not... you may have to wait until things get worse. This is not fun to hear and not fun for us to write, but it is our experience that you cannot manufacture a crisis. You can't force an organization desensitized to failure to feel fundamental surprise. You can't invent a deadline that is truly immovable. You *can* try to increase visibility. We have seen it work once or twice. But most attempts to manufacture crisis conditions backfire and leave you worse off.

We have counseled many people who feel in their *bones* that there's a crisis at hand, yet the situation does not meet any of these criteria. Crisis engineering tactics will not work in a noncrisis environment. You lack all the benefits of a crisis: a deadline forcing new and novel actions, compressed decision-making timelines and hierarchies, and an existential willingness to change.

Our toolkit will get you through most crises most of the time, but not every crisis results in resolution. You may take all the right extraordinary measures and still not succeed. Sometimes the clock runs out. The live event ends. The company goes bankrupt. The product launches and doesn't work. The filing deadline passes. If anyone could guarantee that success was possible, it would hardly be a crisis.

Failures will leave you feeling frustrated, unsatisfied, and unresolved unless you rebracket the story around the learning experience *you* had. What steps can you take to prepare to capitalize on the *next* crisis? How can you become the person the organization will naturally invite to the crisis engineering center or even to run the incident command (if that's what you want)? Perhaps you have long wanted to automate or change a long-standing process, modernize a legacy interface, switch tools, or change the way a key measure is calculated. What changes do you most want to make that you feel blocked on achieving today, and how can you be ready to strike the next time the iron's hot with a ready-to-execute plan and tools?

You may not have made the progress you hoped, or even any progress at all, but you will be wiser for next time.

Section 3:

EMERGING FROM CRISIS

3.1
Know When You're Done

> Occasionally in life, there come times that mark the
> end of puzzles...it is all cockeyed and it all fits.
> —**NORMAN MACLEAN,** *YOUNG MEN AND FIRE*

Running a crisis engineering center as we've described is expensive. People are pulled away from their normal work priorities and lives. They're probably working more hours, more intensely, than ever before. Other organizational efforts are disrupted and pushed aside while the crisis takes priority.

It is important to end crisis mode as soon as feasible. It is better to end the crisis engineering effort slightly early than slightly late.

Some crises end naturally: The merger deadline passes, the concert tour sells out, the power company restores service, the claims backlog shrinks and disappears, or the triage queue at patient intake ends. But if the end of your crisis is less clear, here are some thoughts on why and when to wind down.

ACTIVE CRISIS ENGINEERING IS EXPENSIVE

It is likely that all the individuals actively involved in a crisis engineering effort are experiencing at least some of the following:

- Longer hours
- Hyperfocus
- Lower latencies
- More interruptions
- Higher visibility
- Higher-stakes decisions
- Higher levels of involvement

Intentions and decisions are much more likely to lead to actions in a crisis. This may be an entirely new experience for people used to a molasses-like, risk-diffusing environment.

Leadership is experiencing challenges, too; decisions are now more frequently accompanied by immediate action, along with higher-than-usual visibility.

This is not sustainable. The natural forces and incentives present in all organizations will degrade the crisis engineering center's function over time. The organization will find a way to return to regular order. Anthony Downs calls this the "rigidity cycle":[1] *The moment the crisis engineering effort begins, it is immediately dying and losing its effectiveness.*

The crisis engineering effort will encounter an increasing number of barriers to action as it seeks more resources or authorization. The center will lose its ability to hold the attention of the very authority that convened it to begin with. It will lose its best people, and instead fill with people who do not have answers, nor good questions, and who are there for the wrong reasons. The crisis engineering effort will begin to drift, and eventually sink.[i] Worse, it could be institutionalized in an empty, useless form. It is too expensive to maintain both financially and emotionally.

---

i "Thus, the passage of time weakens the ability of the special organization to retain the concentrated interest of the bureau's topmost officials, which in turn undermines its special privileges. As a result of these factors, the high-productivity phase in the special organization's life gradually comes to an end. It loses its 'special' nature and merely becomes another section of the bureau struggling under the normal weight of rules, regulations, and agonizingly slow decisionmaking procedures" (Downs, *Inside Bureaucracy*, 162).

It is important to bring it to a formal, widely recognized close *before* this happens.

WHEN TO END THE CRISIS RESPONSE

It is nerve-wracking to end a crisis engineering effort. After all, there are still problems to solve! In your likely sleep-deprived state, it may be hard to hold the progress you've made in context. You want more time to do more things in this magical, rapidly closing window of permission and change. But you can't.

This is why we have guidelines on when it's time to shut down the crisis engineering center.

Spin down sooner than feels comfortable.

Slightly too soon keeps the most knowledgeable staff participating through to the end; they will disappear as soon as the effort loses effectiveness. It leaves a better impression on leadership, as the memory will be of a successful effort. It will be much easier to stand up a center next time if the memory of the last one is positive, and if its story is one of effectiveness and success instead of decay and aimlessness.

When you exit, your systems will still be running in degraded mode to some extent. This is normal—they always were, in fact. What's new is a more developed awareness of that degradation.

There will always be more to fix. Systems are always broken in one way or another. If you walk around asking experts what additional work needs to happen, collecting it all in a to-do list, the work will pass on to your great-great-grandchildren.

As a proof point that you can, and should, end a crisis response before every last ember is out, firefighters regularly leave fires *that are still burning* if they otherwise meet exit criteria.[2]

Benchmarks for Normalcy

If your crisis has an inherent benchmark, like "the deadline passed," this is an easy one.

For others, the clearest sign is when the agreed-upon health metrics return to health: The system uptime is at or above goal, the backlog is at or below the volume at which you can confidently clear it by the necessary deadline, etc.

We discussed developing these health benchmarks in chapter 2.7 (Measure Progress). These will vary depending on the organization and nature of the crisis but might include:

- **Stable Workload**—A backlog that is no longer growing and can be managed by the standard team capacity.
- **User Requests**—Inbound user queries and requests returning to historically typical levels.
- **No Critical Escalations**—A set number of days or weeks without urgent customer issues.
- **Operational Stability**—X days or weeks without downtime or system incidents.
- **No Crisis Engineering Center Activations**—X days or weeks without requiring the crisis engineering center's tools or special permissions.
- **Is It Boring?**—Has it been a while since you invented new procedures or deployed new tools to solve problems? Is the agenda the same over and over?
- **Is It Getting Better or Remaining Stable Without Help?**—How long since you invoked executive superpowers or added more personnel? If there are no new actions to take, stop.

Monitoring these benchmarks and beginning turndown as you start to meet them ensures that the organization neither prematurely declares victory nor lingers in crisis mode longer than necessary.

Signs of Morbidity

As another way of looking at it, here are some signs that your crisis engineering effort is nearing death and should be put out of its misery:

- The crisis engineering center has primarily become a means of solving problems unrelated to original crisis circumstances (in other words, the center becomes another "normal" part of the organization).
- People who know what is going on are no longer showing up.
- The crisis engineering center has become a means for achieving visibility (e.g., ladder climbers and attention seekers show up and fan their peacock feathers).

Be bold about hastening turndown if these signs appear.

Watch for the passing of crisis circumstances. It is nearing time to stop when you see less:

- Fundamental surprise
- Failure of sensemaking
- Degradation, disruption, or complete change of core processes or outcomes
- High visibility
- Rigid or nonnegotiable deadline or timeframe

The passing of crisis circumstances will contribute to sponsoring executives losing interest and attention, and otherwise moving on. You want to close up shop before this happens.

Shutting down the crisis engineering center can feel harder than starting it up, especially if the team is exhausted and losing perspective. Use objective health measures, symptoms of morbidity, and enduring indicators of crisis to determine when it's time to end. It is always better to end a little early than a little late.

Next, we'll walk through the steps for wrapping up the crisis effort effectively, and leaving infrastructure behind that you can leverage again in the future.

CRISIS CHEAT SHEET

Active crisis engineering is very costly and will rapidly degrade to diminishing returns as the organization finds a way to return to regular order. Formally end the crisis engineering effort sooner than feels comfortable.

Benchmarks of Normalcy

- Stable workload—backlogs are stable or shrinking
- User requests return to historically typical levels
- No critical problems escalated by users for X days
- No system outages or downtime for X days or weeks
- No crisis engineering center special permissions or tools used for X days or weeks
- Crisis engineering center has become boring or the system is improving without it

Signs of Morbidity

- Crisis team primarily solving problems unrelated to original crisis circumstances
- People who know what is going on are no longer present
- Crisis engineering center becomes a means primarily for gaining visibility

> ### Reduced Crisis Circumstances
>
> - Fundamental surprise
> - Failure of sensemaking
> - Degradation, disruption, or complete change of core processes or outcomes
> - High visibility
> - Rigid deadline or timeframe

3.2

Spin Down the Crisis

> Looking down on the worlds of the Mann Gulch fire for probably the last time, I said to myself, "Now we know, now we know." I kept repeating this line until I recognized that, in the wide world anywhere, "Now we know, now we know" is one of its most beautiful poems.
>
> —**NORMAN MACLEAN,** *YOUNG MEN AND FIRE*

Crisis response demands intensity, focus, and sacrifice from everyone involved. Just as critical as standing up an effective initial crisis response is the process of thoughtfully winding it back down. Done well, this restores a sense of normalcy, acknowledges the efforts of the team, and sets the stage for long-term recovery and change. Here, we outline how to transition from crisis mode back to business as usual, once you've identified it's time.

This chapter focuses specifically on spinning down the crisis response itself; the next chapter focuses on how to instill lasting change, and how to preserve certain new and useful behaviors, tools, and processes.

TURNDOWN

The first step in spinning down a crisis is to deliberately and transparently dismantle the systems, structures, and processes put in place during the emergency. This signals to the organization that the crisis is over and allows everyone to transition back to regular operations. It also keeps anyone from continuing to operate in the shadows.

Generally, the person who initiated a step should be the one who ends it. For example, the original convening authority should be the one to officially decommission the crisis engineering effort, but the incident lead can be the one to disband recurring crisis team meetings.

Turndown is important signaling at every step!

Remove Temporary Approvals and Shortcut Paths

In a crisis, decision-making paths are often streamlined to respond quickly to pressing issues. However, continuing to bypass standard processes undermines long-term stability. Instead:

- **Remove Temporary Approvals.** The original convening authority needs to revoke any special permissions granted for rapid decision-making during the crisis. Reinstate regular approval chains.
- **Communicate Changes.** Ensure everyone understands that the shortcuts have been retired and normal procedures are back in effect.
- **Save What Works.** Institutionalize new decision-making pathways that are high functioning: those that are fast, and more accurate than not.

End the Crisis Engineering Tempo Check-Ins

During the crisis, frequent check-ins maintain alignment and urgency. As the situation stabilizes:

- **Disband the Standing Crisis Meetings.** Clearly announce the end of crisis-specific check-ins and meetings. You may need to declare a phase-out plan, e.g., the hourly meeting will move to daily for the next week or until we hit a certain milestone, and then end at that point.
    - You need to *actively* disband each meeting, or you'll discover this time next year that people are still showing up to an infinitely recurring calendar invite. (We are *still* on calendar invites from clients from years ago.)

Reintegrate Teams

Crisis response pulls individuals away from their regular roles and teams. The original convening authority should:

- **Send People Back.** Officially reassign team members back to their original teams, roles, desks, and schedules.
- **Express Gratitude.** Acknowledge their contributions and ensure their original teams are prepared to reintegrate them smoothly. It's bad if a valuable team member finds themselves unable to return to their prior role, and subsequently regrets participating in the crisis effort.

Turn Off Temporary Infrastructure

During a crisis, new tools, platforms, and workflows are often created to address specific needs. As the crisis resolves:

- **Decommission Crisis Tools.** Turn off temporary infrastructure like dedicated phone bridges, shared whiteboards, and crisis-specific dashboards.
- **Archive Critical Information.** Ensure that any important data from these tools is saved and stored for future reference (in addition to adhering to any organization-specific retention policies that may have been overlooked in the haste to get set up).
- **Save What Works.** Preserve and promote new, high-value communication forums and dashboards where they will remain relevant.

Turning this infrastructure off is one of the clearest possible signals that the crisis response has ended.

Shut Down the Crisis Engineering Center

Publicly and formally disbanding the crisis engineering center and the staff working in it sends a physical message to the organization that it is returning to regular order. The convening authority who set it up should be the one to announce its end.

Hold a Closing Ceremony

Marking the end of a crisis with a ceremony or ritual also provides clear closure. Options include:

- A formal announcement from leadership, praising the efforts and the outcomes
- A brief all-hands meeting to reflect on the response and acknowledge the team's efforts
- An announcement via the status page, all-hands email, and/or displays in the lobbies of any buildings

You may even make a big deal out of turning off the temporary infrastructure, like unplugging a screen or clicking Delete on a phone bridge or Teams room in front of an audience.

AFTERCARE

The conclusion of a crisis is a crucial time to care for the team and address the lasting impacts of the intense experience they just had. Done right, this will create a resilient crisis response force for the future of the organization. If the crisis team flames out in a tornado of burnout and disillusionment, good luck getting even B-team players to join a future response.

It's important to understand that the strain and the novelty of the crisis experience will affect team members' health, relationships, and personal lives. Some may even acquire post-traumatic stress disorder. We can't begin to address mental health in the space we have in this book. Seek professional help in any cases where mood and behavior changes are interfering with a person's life. Do not delay.

Here are some actions that you can take as an incident lead that will help to heal the team and its members.

Encourage and Enforce Vacations

Team members will be exhausted after the crisis. This is bad for their health (physical and mental) as well as their future performance. A crisis that did not feel particularly stressful for one person may have been the most stressful experience of another member's entire life.[1]

- **Recommend Time Off.** Encourage individuals to take breaks and recover. If necessary, you may have to require some people to take time off in order to prevent burnout. Leaders should model this behavior by taking time off themselves and *meaning it* (e.g., not answering emails or jumping on calls while off). Reinforce this with photos and stories of no work happening on vacation.

Watch for New Aspirations or Disillusionment

Crisis experiences can profoundly impact individuals' perspectives and career goals. Exposure to the breadth of how an organization really works can cause people to feel either newly empowered or disillusioned.[2] Be attuned to:

- **New Aspirations**—Some team members may discover new skills or ambitions. Expect some to want a role change or new team.
- **Disillusionment**—Others may feel disengaged or ready for a change due to the intensity of the experience. This may mean itchy feet to leave the organization entirely, or to take on a substantially different role. It can be hard to feel enthusiastic about business as usual after a singular experience. (Ask Frodo!)
- **Comedown**—Some team members may feel depressed by a return to "regular life" and regular-order responsibilities.

Expect changes in vendor relationships or team structures as well. The crisis likely revealed gaps or opportunities that can result in a musical chairs of personnel.

None of these are inherently negative, but if you pay attention, you can keep them from blindsiding management or disrupting the organization.

For example, if a key employee now has itchy feet, you may have to do some serious negotiating to keep them around long enough to backfill them. And while someone may hunger for more responsibility, being part of the crisis response may not necessarily have fully prepared them for the bigger role they're now after.

Create a Memento

A tangible reminder of the team's efforts can be a meaningful way to celebrate their contributions and reinforce the new social connections crisis team members have forged. Ideas include:

- Stickers, mission patches, or challenge coins
- A printed or digital keepsake, such as a timeline poster, yearbook, or photo album
- A team dinner or an event to reconnect and reflect

Share the Story

Capture the lessons learned and share them with the broader organization. Options include:

- Writing an after-action report detailing the crisis, the response, and recommendations for the future. This works best when it's easily accessible by anyone interested (and by future organizational archaeologists).
- Giving a presentation or talk to share insights and acknowledge contributions.

Plan for a Reunion

For significant crisis response efforts, consider organizing a reunion event, such as a dinner or a small gathering, to mark the anniversary of the incident. This can provide a positive and reflective touchstone for the team. The HealthCare.gov rescue team continues to hold reunions more than a decade after the initial response.

Winding down a crisis requires thoughtful planning, clear communication, and care for the individuals who gave their all to respond. Taking deliberate steps to deconstruct the temporarily constructed systems, acknowledge contributions, and allow critical contributors to recharge is an important part of the process.

> ## CRISIS CHEAT SHEET
>
> Turning down the crisis engineering effort intentionally, formally, and methodically is critical. It must be done actively.
>
> ### Turndown
>
> - Remove temporary approvals or special permissions, communicate those changes, and specify which normal procedures are back in effect, while saving the new ones that will continue to serve you.
> - End the crisis engineering tempo check-ins, *actively* disbanding all standing crisis engineering meetings.
> - Reintegrate personnel, officially reassigning crisis team members back to their original teams, expressing loud and public gratitude to the personnel and the teams.
> - Turn off temporary infrastructure like phone bridges, chat channels, shared whiteboards, and crisis-specific dashboards, archiving critical information and saving what new communication forums and dashboards remain relevant to regular order.
> - Shut down the crisis engineering center, take the sign off the door, and return the space to normal use.

- Hold a closing ceremony, including a wide announcement and ceremonial actions held in front of a broad audience.

Aftercare

- Recommend or enforce time off for individuals intensely involved in the crisis engineering effort.
- Watch for new aspirations among team members who have developed new skills and/or ambitions.
- Watch for disillusionment among team members who may feel disengaged after the intense experience.
- Watch for comedown among team members depressed about a return to "regular life."

Preserve the Story and Its Tellers

- Create a memento for the team.
- Share the story by capturing and telling the lessons learned repeatedly to internal and external audiences (as appropriate).
- Plan for a reunion of team members on the anniversary of turndown.

3.3

Instill Change in the Wake of Crisis

> The Mann Gulch tragedy immediately became a flaming symbol to the Smokejumpers and to firefighters generally, especially those in the Northwest. Fortunately, there are a lot of able woodsmen in the Forest Service who don't wait around for the Forest Service to do something, and it was some of these who said to me not long after the fire, "God damn it, no man of mine is ever going to die that way."
>
> —NORMAN MACLEAN, *YOUNG MEN AND FIRE*

If you've read this far, you know that a crisis is the best time, and often the only time, to make permanent lasting change in your organization. The Jedi mind trick is to know this going *in*, so that you can steer in the direction of the changes you want, and actively continue the new practices that you want to keep (while disposing of the old practices that you don't).

You need to be quick, since the organization will revert to business as usual in a few days or weeks at most. If you take too long to decide or to take action, this inevitability will beat you to the punch.

A crisis can't change everything, so you need to be selective and strategic. We'll share some advice on how to strike this balance.

In this chapter, we'll also point out some useful behaviors you may want to preserve postcrisis, and how to adjust them to fit into an everyday world. Temporary infrastructure and processes developed during the crisis are a reference implementation for what the organization needs in the long term. And if there are practices you don't want to keep, you'll have to *actively* end them, or they'll continue on as disruptive ghosts for years to come.

While we encourage focusing on big transformative goals, there's usually also a handful of smaller behaviors and artifacts that can persevere postcrisis with the right shepherd. It's worth running through this list to make sure you don't miss any.

USE THE RIGIDITY CYCLE TO YOUR ADVANTAGE

The rigidity cycle observes that an organization always moves toward hardening, becoming incapable of fast or novel action as it "slowly and inflexibly grinds along in the direction in which it was initially aimed."[1] It dictates that the moment the crisis engineering effort begins, it is dying. The organization wants to, and will, return to a state of stability.

You cannot stop this from happening. Many people need to learn this the hard way.

Crisis engineers embrace the rigidity cycle instead of fighting it, quickly arranging the organization into the new desired position before it inevitably rehardens, and taking care that unwanted behavior does not stick around long enough to fossilize.

INSTILLING CHANGE IN THE WAKE OF A CRISIS

If you act quickly, you can permanently stop, start, and change some organizational behaviors for the better. If you ignore this moment, habits will likely change anyway, in ways you may not like or be able to predict.

A crisis gives you a possibly once-in-a-lifetime opportunity to demonstrate that a change you've wanted to make for a long time—reorganizing the department, adding automated tests, ending a painful process step, using a more creative definition—is actually better and less risky than the status quo. In peaceful times, it may be impossible to get permission, or to create the circumstances to demonstrate the superiority of your approach.

You want to be thinking about this from minute one of the crisis. (A good crisis engineer is *always* planning ahead for how to take advantage of a future crisis.) What's the change you want? What change do you expect it to make? What has been the pushback so far? If crisis conditions create an opportunity to stop/change/start, can you take advantage? Are there people you should make sure can see and understand the change and its impact early on? Is there data you could collect to make your case? Take action, watch what happens, and document/expose the results to the extent practicable.

As difficult as it is, you need to remain open to the possibility that your change did not create the improvements you expected, that it actually made things worse, or, more likely, that it created improvements in one area but unexpected consequences in another. Once you know this, it can shape your postcrisis efforts to make it permanent (or not).

What Do We Even Mean by Transformative Changes?

It's our experience that people interested in crisis engineering already have laundry lists of changes of all kinds and magnitude that they would like to make in their environment. When we show up on a job, we can quickly come up with some ways that client can emerge stronger and better from the crisis at hand—not because we're psychic, but because a chorus of people point out the same barriers or broken processes to us from day one. But if you need some ideas for a transformative change, common ones include:

- Centralizing or decentralizing the IT department or another enterprise function
- Enforcing improved, multifactor authentication in the wake of a hack
- Automating a process
- Transforming/modernizing/digitizing a process
- Changing a practice
- Adopting a new best practice or standard of care
- Measuring/improving the user experience
- Automating tests or implementing "DevOps"
- Using commercial cloud
- Practicing at the top of your license
- Hiring a new team
- Turkey-farming a current team
- Speeding up
- Slowing down
- New measurements
- New goals or priorities, or deprecating a prior goal/priority

Can You End the Next-Generation Project?

It's rare for us to come across an organization that does not have a "modernization" project in progress that has been planning Big Things for going on a decade, with the promised land only "18–24 months" from today. This timeline never changes. In the meantime, the best talent is pulled away to work on it, and all but the most urgent changes are postponed while we wait for the new world.

Leadership often believes that if the next-gen, multiyear modernization project had already launched, they could have avoided this crisis. But it didn't, because here we are. If parts of that long-term strategic effort were ready and agile enough to be helpful mitigations, then keep them. The more common outcome is that the next-gen project should be canceled and reimagined based on the real-world events that just happened, and the demonstrative changes made to the existing system. Maybe people have been suspecting it was doomed for a while—now you have the opportunity to try again, with a series of changes to the existing system instead of a decade-long planning process.

Can You Permanently Stop Doing What You Temporarily Stopped Doing?

Complex systems are strengthened by subtraction.[2]

A pattern we see often is a pile of instrumentation, process, and workarounds, each tracing its heritage back to a prior incident. These will coexist and conflict with the normal way of doing things. This is a brief opportunity to ruthlessly prioritize, reorganize teams, and shut down projects based on what you've now learned really matters.

If you have run your crisis team successfully by skipping some process, consider removing it entirely. This team should be seen as an example of how to accomplish things quickly and safely.

CRISIS BEHAVIORS TO PRESERVE

The worst possible postcrisis behavior is divergence, where some teams keep a random assortment of new practices and others go back to the previous normal. Now you have a greater amount of process that's even more disjointed. Exiting a crisis with twice as much process and infrastructure is asking for another crisis.

But you can't keep everything. So, what's worth preserving?

Do the Same Thing but Slower and More Rigorously

Crisis response requires urgency and acceleration of normal controls. It also tends to set up high bandwidth and frequent communication. Many new practices can be slowed down to a sustainable cadence and formalized to be effective in the new normal. Doing this also makes it easy to respond to a future crisis, since you know those things can be sped back up. This will work in part because the processes themselves are comprised of either different staff or similar staff with different knowledge than before the crisis. Some examples:

- Twice-daily triage of key customer issues during a crisis can become a weekly effort
- Nightly system patching with verbal sign-off during a security incident can become a monthly process with tracking and internal notifications
- Immediate approval of firewall changes can revert to a weekly update attached to an existing team sync, instead of the prior three-month wait

Formalize Ad Hoc Tools and Processes

A crisis can expose holes in existing tools and processes. People will fill these holes with spreadsheets passed around via email and scripts maintained outside normal development cycles. These human workarounds are helpful to resolve a crisis, but when the dust settles, the organization needs to prioritize formalizing (and where possible automating) these shadow processes. If these processes remain manual and undocumented, they risk causing a brand-new crisis, or at least making the next crisis harder to investigate.

One way to think about it is that these ad hoc adaptations are basically a list of changes the system needs. The more widespread and/or difficult an adaptation is, the higher priority it should be to formalize.

If you temporarily paused or modified a policy for the crisis, can you permanently codify that change? This may involve expanding eligibility criteria, reducing the number of sign-offs, converting a serial approval process to a parallel one, or suspending a burdensome requirement that may not be all that necessary.

New, useful practices, like bringing together decision-makers from different areas, generating a new report, providing a plain-language translation to a highly technical update, or having a technical writer sit in the crisis engineering center to update documentation in real time, are worth noting and actively either preserving or discarding.

We appreciate that in some environments, especially more regulated ones, this formalization may take significant time and effort. That's okay!

Look for Capabilities That Can Be Useful in the Future

Among the new tactics invented during the crisis, look for ones that can be useful in future situations. These may include blocking certain users, skipping parts of a process pipeline, or doing outbound customer support (calling/emailing your customers proactively). Document these mitigations and keep them in use out in the open, not in the shadows.[3]

Avoid Restarting Technical Changes from Scratch

It is tempting to discard technical changes made during a crisis as "duct tape" or technical debt. Some truly temporary changes, like rate-limiting free users of the system, should be reverted, but most should be rolled forward into supported parts of the infrastructure.

Don't start over with a fresh team and reimagine the requirements through your normal process. That will almost certainly give you a second-system effect.[i, 4] Instead, take the "hacked" code, and some of the same people, and evolve the "duct tape" to meet your engineering best practices. Again, the adaptations are the requirements.

Codify Key Health Metrics

The small (less than ten) number of end-to-end metrics used by the crisis response team should be kept as the organization-wide Key Performance Indicators (KPIs). They are likely to be higher fidelity and more holistic than whatever existed before, or you would have caught these problems before they became a crisis.

Maintain Some Artifacts Created During the Crisis

Some sensemaking artifacts should be preserved and used as the basis for ongoing documentation and planning. These include process maps, system diagrams, and organizational charts, as well as lists of priorities. Documents created during a crisis are more representative of real-world conditions than those from the start of the project, formalized regulatory documentation, or aspirational plans.

For technology and process practices, keep those that are directionally aligned with where the organization wants to be.

If part of the crisis response was propping up a soon-to-be turned-down business line, take the opportunity to accelerate that turndown. If the crisis response has set up a high-touch customer support process and that is an area where the company is investing, keep it!

---

i This is where an overly confident team, on the heels of successfully launching a simple initial system, will build a bloated, expensive failure of a system the second time, which wildly overcompensates for anything the first system lacked.

ACTIVELY END UNWANTED CRISIS BEHAVIORS

Don't let unwanted behaviors stick. New crisis-induced behaviors, whether written down or not, need to be identified and intentionally stopped or modified before they go on for too long, or they will become part of your organization's new way of operating. This is the origin story of many a "water cooler rule," where an organization does something ceremonially for decades without ever understanding or questioning why.

While working on HealthCare.gov, in the height of the crisis, Mikey implemented a nightly 3 a.m. phone call to a senior White House official to provide a status update. This made sense at the time, because the White House wanted to know the latest before the country woke up and the news cycle harped on any new issues. When Mikey returned ten years later, he learned that this 3 a.m. phone call continued to happen every night. When he pointed out that he was the one who made that rule in the first place and that it no longer made sense, no one cared. He had no authority to change the "standard operating procedure manual." This pointless phone call continues to this day.

USE THE PEOPLE FROM THE CRISIS EFFECTIVELY: THEY'VE SEEN SOME SHIT

The individuals who navigated the crisis can be some of your most valuable assets moving forward. They've developed rare experience, built trust across siloed teams, and gained insights that can strengthen the entire organization. You can intentionally harness and support their growth:

- **Nurture New Social Connections.** Our recommended approach to crisis engineering is building a small, multidisciplinary team with a steady operational cadence. Individuals who participate in the response

are ideally suited for seeding new teams that work in an improved style. These people can also be encouraged to act as organizational guides and connectors between teams.
- **Develop New Skill Sets.** You now have a new set of individuals who understand how to run a crisis engineering effort. Keep these team members engaged. Occasional social events or even ad hoc opportunities to bring the gang back together again to weigh in on a new idea can help keep relationships warm, which in turn can keep disparate parts of the organization in contact and collaboration in ways they otherwise would not be.
- **Elevate the Gap Fillers.** A crisis will often reveal individuals who have been patching over systemic failures through sheer force of will for years. The ones who route the tickets in the support queue by hand so critical customers get faster service, or have a library of database queries to fix up common data problems. They are doing these things in addition to whatever their job title says. At minimum, you should make sure they get appropriate credit for this work. They should also be treated as experts to guide future system and process development.

ADDITIONAL THINGS TO CONSIDER PRESERVING

Don't miss an opportunity to keep these artifacts around while you're at it:

- **Better and Real-Time Data**—If you're generating more accurate, more timely, and/or more relevant reporting…keep that up! Are there members of the data team who could be better looped into ongoing meetings and measurement efforts? If you upgraded dashboards or data connections in a few places, can you now continue that work more widely?

- **Clear, Finite Priorities**—The crisis likely crystallized the organization's goals and how to measure them. Great. Highlight and keep key health metrics front and center. If your organization has historically had one thousand dashboards, this is a good time to pick the top three and turn off the others.
- **Clear(er) Decision-Making Processes**—As part of how you ended up in crisis in the first place, you may have identified unclear decision-making processes, or decisions made without understanding the downstream impact on other parts of the organization. Developing effective decision-making processes is the topic of decades of organizational behavior research, but a crisis is certainly a time to keep identified improvements.
- **Backlog/Task List**—Many tasks were likely identified in the crisis response that are not yet finished, or even started. Those should live somewhere to ensure the important ones (not every last one) are evaluated and completed.
- **Cross-Organization Communication**—Parts of the organization that have rarely or never met have now been in close communication. This is expensive to do frequently, but should be encouraged to happen occasionally. As you emerge from the crisis, nudge team members to consider whether newfound teammates should be added to key brainstorming, strategic, or resourcing efforts. If nothing else, making some petty cash available for cross-functional lunches can be valuable.
- **Status Reporting and Updating**—You don't want the organization to drown in alert fatigue, but if you stood

up a status page or notification list, that's a resource you can continue to use to disseminate accurate information. If you developed new muscles and habits around communications, like separating out highly technical updates and providing plain-language explanations, or inviting technical writers to join your deploy process instead of hearing about it after the fact, keep doing that.

- **Improved Maintenance Resources**—We often find a situation where infrastructure collapsed after many years of underresourcing (or flat-out not resourcing) maintenance tasks.[ii] This can include technical maintenance like upgrades and bug fixes, as well as process improvements like updating documentation, creating user-friendly content, and updating job descriptions and performance plans. This is the window to advocate for the time and funding resources that are needed for a permanent maintenance budget.

Emerging from a crisis is not simply about survival—it is a rare and powerful opportunity to shape the future of your organization with intention. By acting swiftly, preserving the right adaptations, and decommissioning what no longer serves, you can guide your organization to a stronger, more resilient state rather than letting chaos or inertia dictate its course. The lessons, tools, connections, and clarity forged under pressure are some of the most valuable assets your team will ever create. Seize this moment thoughtfully, and you can lay the foundation for enduring excellence—before the inevitable rigidity of normal order sets back in.

[ii] Deb Chachra, in her book *How Infrastructure Works* (Riverhead Books, 2023), calls this deferred/ignored maintenance, or underinvestment, "red termite risk."

CRISIS CHEAT SHEET

The moment a crisis engineering effort begins, its ability to accomplish new things begins to diminish. The organization will always find a way back to regular order, per the rigidity cycle. You must plan ahead on which changes to preserve in the new, postcrisis regular order.

Instilling Change in the Wake of a Crisis

- Look for transformative changes: reorganization of functions, transforming entire processes, hiring a new team or sidelining an existing one, speeding up or slowing down, creating new priorities or formally deprecating old ones.
- Consider bringing an end to stalled or failed multiyear "modernization" projects.
- Look to permanently stop any processes, instrumentation, or projects that were temporarily suspended.

Crisis Behaviors to Preserve

- Do the same thing but slower and more rigorously.
- Formalize ad hoc tools and processes.
- Look for capabilities that can be useful in the future.
- Avoid restarting technical changes from scratch.
- Codify key health metrics.
- Preserve some of the more accurate artifacts created during the crisis.

Actively end unwanted crisis behaviors before they are informally enshrined as cultural "rules."

Use the People from the Crisis Effectively

- Nurture new social connections across the organization.
- Develop new skill sets by keeping crisis engineering team members engaged on decisions.
- Elevate the people you've discovered who close gaps in the system.

Additional Things to Consider Preserving

- Better and real-time data
- Clear, finite priorities
- Clearer decision-making processes
- New backlog/task list definitions
- Cross-organization communication forums
- Status reporting and updating
- Improved maintenance resources

3.4

Plan for a Future Crisis

> How do we get from where we are to where we want to be, without being struck by disaster along the way?
>
> —PAUL H. NITZE

The number one question we get asked in our crisis engineering workshops is: "How do I prevent a crisis?"

You cannot prevent all crises. Indeed, if you've read this far, then you understand that what you are actually asking with that question is: "How do I prevent my organization from ever making fundamental changes or adaptations to new circumstances?"

However, this doesn't mean you should be blindsided by something new (or something old) every morning. Learning about your system over time should expand the universe of foreseeable events, making fundamental surprises (and thus crises) less frequent.

Building your crisis engineering capacity, as described in this chapter, can certainly help you respond to and leverage future crises more effectively. You can assemble the team ahead of high-risk launches and deadlines.

Implementing good "crisis hygiene," developing distortion-resistant reporting that reduces surprises, and practicing operating on multiple time scales can all help you to be better prepared for your next crisis. Finally, a skilled crisis engineer will always have transformative ideas ready to go when the next crisis presents itself.

BUILD YOUR CRISIS ENGINEERING CAPACITY

An organization with the ability to spin up a crisis engineering center on an arbitrary topic at any time will be able to jump on even small issues without hesitation.

Read through chapter 2.2 (Establish a Crisis Engineering Center) in calm times and consider how you would implement these steps today. For example:

- What room would serve as the crisis engineering center? Does it have whiteboards, chairs, and internet?
- Is there an organization-wide phone bridge with a well-known number?
- Does the organization have a status page that people know about and rely on?
- Does the person you'd expect to be the convening authority know what a crisis engineering center is?
- Is there a "bench" (list) of possible crisis response leads?

If you recently had a crisis and followed some of our advice, look back on what happened, what you wish you had done differently, and what tools or resources you wish had been available. Work on removing those roadblocks or procuring those tools. For example, an always-on channel for reaching the crisis engineering team is a critical component of a crisis engineering center. If you couldn't stand up exactly what you wanted during this last crisis, work on getting one set up before your next one.

You should also package up any available resources, such as communications templates and the crisis announcement. A particularly motivated organization could develop some of these materials in advance.

Of course, deepening the organization's understanding of crisis engineering by distributing this book, reading complementary materials like Karl E. Weick's book *Sensemaking in Organizations*, and hosting crisis engineering workshops and training can take your readiness from simply having the room and the status page on standby, to being ready to harness the opportunity of the crisis.

PREPARE FOR KNOWN SPICY DAYS

Many organizations have risky events, where a surge in users, a key deadline, or a system or rule change make the stakes higher (and the number of people impacted greater) if something goes wrong. Common examples are:

- E-commerce over the holidays
- The Super Bowl
- Enrollment periods
- Elections or tax filing deadlines
- Big public announcement (e.g., a stock offering, a merger, or an annual keynote)
- System upgrades/rollouts/launches

You might also face what's known as "gray swan" events, which are events that you definitely expect to happen, though you don't know when.[1] The next Carrington Event (an intense solar storm that last happened in 1859) is one example; one might argue that a wildfire in California is another. These may have direct impacts on your line of work.

For whatever these are in your organization, get the team together ahead of time.

1. Set up your crisis engineering center and bring the team together. In a planned scenario, it should be even easier for everyone to know which room to go to and which conference bridge to dial into.

2. Give them space and permission. You may need to reinstate expedited decision or approval processes in advance of any issues. The team has to be focused; their other roles and responsibilities need to wait until after the response is over, even if no crisis appears this time around.
3. Check in on, and strengthen, prior crisis behaviors. What went well? What went awry? This is the time to practice wobbly processes and to close previous gaps. Just because your last effort went well doesn't mean this one will go smoothly. Keep your eyes open for opportunities for improvement, and for variations from prior efforts.
4. Create crisis engineering internships. A known crisis is a chance to bring in one or two (but no more than that) promising new team members to expose them to the crisis engineering efforts and build their skill set.

CRISIS PREVENTION HYGIENE

Some practices will inherently make it more likely for you to keep some crises from ever happening. These include:

- Defining and monitoring key system health metrics, with alerts (not too many!) to people who can do something about them before catastrophe strikes. These might include network latency, backlog size, or response time.
- Practicing good IT behaviors like regularly backing up systems, successfully restoring from those backups on a regular basis, and not storing the backups on the servers being backed up.
- Building in resiliency and redundancy that makes sense for your organization. If you believe you already have redundancy or fallbacks built into your system, how are you testing this?[i]

---

[i] The infamous O-rings that caused the *Challenger* to explode *were* a redundancy—but an untested one. It can take some creativity to truly test your failover.

- Simulating your most likely scenarios before high-priority events/deadlines or significant technical or policy decisions. Bring all the relevant players together in a room and act out different scenarios to identify and ameliorate gaps, inconsistencies, and unmapped territory.
- Not delaying regular actions. *Sensemaking requires action.* The longer your organization spends thinking about a problem or possible solution, the further it drifts from reality. Look for opportunities to act. Even smaller actions like prototyping, piloting, writing a test script, or crunching the data on an assumption help fight inertia.
- Maintaining your system. The biggest infrastructure collapses in technical systems, and the built world, are often attributable to corrosion and degradation over time.[ii] As Deb Chachra says, "Any sufficiently advanced negligence is indistinguishable from malice."[2]
- Testing your crisis communications channels. This is in part to ensure you can get messages out, but also to ensure these channels can't be accessed in a non-emergency by mistake, which can create its own kind of crisis.[iii]
- Building for flexibility. This can be as simple as using a variable instead of a hard-coded value in a script so it can be used by other teams, or, when updating a mission-critical system to comply with new regulations, considering how to better accommodate when that regulation changes again in the future.

---

ii And squirrels, in the case of the built world.
iii See the false alarm in Hawaii that texted everyone they were going to die.

IMPLEMENT DISTORTION-RESISTANT REPORTING

In many crises, warning signs were available long before disaster struck—but they were buried under layers of distorted reporting. Metrics that show only good news, executive summaries that sand off rough edges, and cheery PowerPoints that mask brewing failures can all contribute to a false sense of security. Sadly, this is how most executives get their information: through a chain of updates that tweaks and sanitizes data through many levels. Any real data in these reports inevitably becomes stale, due to the length of time this process takes.

To prevent this, you must create reporting mechanisms that resist distortion at every level.[3] This means building dashboards and reports that show unfiltered reality, even when it's uncomfortable. Reports should be automatically generated as much as possible, based directly on operational data rather than handmade summaries. Establish clear and widely understood definitions for metrics, including what constitutes a good or bad status.

Leaders must encourage a culture where bad news can surface early without punishment. They should also model curiosity about surprising or negative results instead of demanding only positive outcomes. A team that believes it can safely report problems is one that can catch small fires before they become infernos.

Last, periodically audit any reporting pathways you rely on: Check that frontline observations are making it to decision-makers without being overly summarized, sanitized, or delayed. Anthony Downs suggested sampling reports randomly, not necessarily prioritizing the most important ones,[4] so the organization cannot adapt its distortions. Crisis prevention depends not just on having information, but on having *real* information, in time to act on it.

> Are you changing the story or are you preventing a future crisis?
>
> It's tempting to redefine a term or metric when it's trending in the wrong direction, instead of actually taking action to change reality. Take care that efforts to prevent a crisis aren't misapplied to suppress concerning information.
>
> For example, the National Highway Traffic Safety Administration banned the word "accident" from government publications in 1997.[5] That's certainly one way to reduce accidents, but not one with any actual impact on safety.

PRACTICE OPERATING ON MULTIPLE TIME SCALES

We talked earlier about complexity at different sizes of system, and when you should zoom in or zoom out. When you vary the time scale, you'll also expose different levels of complexity.

Consider Three Mile Island. There were many possible actions available to the operators in the control room; they were visible as switches and buttons. The lag on these control systems was seconds to minutes. There were several points in the accident sequence where operators could have prevented the accident, but they had only a minute or two to find the correct action. It didn't matter if the solution was discoverable by a committee of nuclear physicists or documented in a thousand-page manual. Nothing but trained intuition could have saved the reactor from inside the control room.

Imagine if Karl Weick, Anthony Downs, and Stafford Beer had all been standing in that room. It's very unlikely it would have done anyone any good. There wasn't enough time or decision space.

On the other hand, imagine if they had been looking at the information flows between the reactor operators and the manufacturer.

Without understanding a word of the accident report, they might well have realized that the 1977 incident at Davis-Besse was important information for the other operators. Group sensemaking across the reactor fleet could have prevented the accident.

The control loop by which new reactor designs are built and tested runs even slower—so slowly, in fact, that it produced not a single new license between 1978 and 2023. The total stall in sensemaking at this layer is the reason that 1971 designs were in use (like many other places) at Fukushima Daiichi (the site of another nuclear accident) in 2011. Better designs were known before that.

Improving the sensemaking of any of these control loops could have prevented the accident at Three Mile Island. Designers could have made the accident impossible decades ago. The manufacturer could have corrected the misunderstandings months before the accident. The operators could have averted disaster in the final few minutes. There's no reason to think one loop is more important or valuable than the others. They should all be on the table if you are conducting a project to diagnose or improve the complex system.

This means you need to become comfortable thinking and operating at different clock speeds. Most likely, a person interested in our "crisis engineering" pitch is already good at one tempo. Consider practicing the others. If you spend most of your job setting up quarterly plans for data center expansion, try visiting the people who replace hard drives every day. If you normally adjudicate exceptions on individual applications for unemployment benefits, try visiting a policy meeting.[iv] A single person who knows about several layers will come up with better optimizations than a committee of people who each know about one.

---

iv We know: One of these directions will be easy and the other will be hard. For some reason, decisions that take a long time to implement are perceived as "strategic" and done by higher-status people. We don't have a good theory as to why. It is obviously not because they add greater value to the organization.

HAVE YOUR IDEAS READY FOR THE NEXT CRISIS

In the compressed decision-making timeframes of a crisis, the organization can only consider a few options, relative to how many it could consider with a multiyear committee. If your option is ready to go, it has a good shot of being selected in a crisis even if it wouldn't have been *considered* in other times.

What transformations and changes have you been unsuccessful at getting traction for? (Peek back at the list in the previous chapter for help brainstorming.) Do you know what would have to change in your environment to make your proposal more attractive (for example, a changed priority, a different calculation of a metric, a sudden inability to hire more people on a task)? Do you have a plausible first and next step? Have you written a one-pager?

Some crises are inevitable, but with thoughtful preparation, distortion-resistant reporting, and deliberate practice, you can dramatically reduce their frequency and severity. Building and maintaining these habits can create an organization that is not just resilient in the face of disruption, but that is capable of adapting, learning, and emerging stronger each time. You can't prevent every crisis, but you can take steps to prevent some of them.

CRISIS CHEAT SHEET

You can't prevent all crises, and if you did, you'd prevent the organization from ever making fundamental changes or adaptations to new circumstances.

Build Your Crisis Engineering Capacity

- Consider the steps to establish a crisis engineering center, and how you'd implement them today if necessary.
- Package up any available resources like communications templates and the crisis engineering announcement, or develop them in advance.
- Deepen the organization's understanding of sensemaking by reading related materials and holding crisis engineering workshops.

Prepare for Known Spicy Days

- Launch dates, deadlines, enrollment periods, live events, gray swans—some days on the calendar are more likely to have a crisis than others.
- Set up the crisis engineering center in advance.
- Bring the team together, and give them the necessary space and permissions to act if things go awry.
- Check in on, and strengthen, your prior crisis behaviors.
- Create one or two crisis engineering internships to help more people learn the ropes.

Crisis Prevention Hygiene

- Define and monitor system health metrics, like uptime or backlog size, and set alerts.
- Practice good IT hygiene like backing up systems, restoring from backup, and failovers.
- Build in resiliency and redundancy that makes sense for the organization, or test it if you think you already have it.
- Simulate your most likely scenarios before high-priority events or deadlines.
- Do not delay regular actions. The more time spent thinking about action, the further the understanding of the action drifts from reality.
- Maintain your system; time erodes all solutions.
- Test your crisis communications channels.
- Build your flexibility.

Implement Distortion-Resistant Reporting

- Build dashboards that display unfiltered reality, even if it's uncomfortable.
- Automatically generate reports, without human meddling or manipulated summaries.
- Establish clear and widely understood definitions of metrics, what is good or bad.
- Audit reporting pathways to ensure frontline observations are making it to decision-makers without too much summarization, sanitization, or delay.

Practice operating on multiple time scales at once and have your ideas ready for the next crisis.

3.5

Engineer a Crisis Career

> He who fights with monsters should be careful lest he thereby become a monster. And if thou gaze long into an abyss, the abyss will also gaze into thee.
>
> —**FRIEDRICH W. NIETZSCHE**

The human parts of a complex system accept meaningful change only during certain short time windows. One of the reasons this idea isn't more popular is that it's difficult to work into a career plan.

Making meaningful change in a complex system isn't necessary for the careers most people want. Few people would acknowledge this about themselves. For some reason we live in a culture that mass-produces fresh-faced college graduates who believe they must now "make a difference," despite having no idea what that difference is.[i] But if you ignore words and look at actions, it's clear that most of the time most people do not want to be change agents. They want stability, money, a relatively peaceful workday, power, and the time to do other things that they care about. When a crisis arrives, they step back and wait for a change agent to appear.[ii]

---

i What *do* you do with a BA in English?
ii Even Mr. Rogers encourages us to wait, with his famous call to "look for the helpers" in a crisis.

This is okay. In all organizations past the startup phase, the routine actions that need to be repeated in order for an organization to exist tomorrow consume the vast majority of the energy. On the most intense day of the most existential crisis, that overhead is *still* most of what's going on. The crisis engineering center stops being habitable pretty fast if no one empties the trash or the Wi-Fi goes out.

Another argument that this must be okay is a proof by existence: It's how things are, and the sun keeps rising.

If you made it this far, we assume you want to make changes in your own career. It's likely that, at least, you carry some dissatisfaction with some aspect of how the complex system around you behaves. You have some appetite for a departure from normal order, for an increase in novelty and change. We'll also assume that you want to sustain some number of expensive habits like eating and sleeping indoors, so you will need something approximately like a job and a paycheck.

BROADEN YOUR HORIZONS

Most of the time, a complex system will tolerate only slow, incremental change. Opportunities for radical change are sporadic and unpredictable. It follows that the best way to increase your exposure to opportunities for change is to have a connection to multiple complex systems.

There are several ways to interpret this idea. If your current role limits you to working on one system, look again. You can probably zoom in and find subsystems that operate with some amount of autonomy, and that experience their own minicrises. The overall order fulfillment system may be frozen in amber, but there is probably some part of it that is such a mess that anyone would be glad for someone else to burn it down. Individual contributor jobs in engineering or operations are usually embedded in this kind of structure.

Middle managers and executives are already in the position of overseeing multiple systems. It's curious that they almost always see them as multiple systems, rather than as one integrated whole. This is probably a cousin of **Conway's Law**, attributed to Melvin Conway in 1967, which says that organizations can only design systems that mimic their own communications structure. We would restate this as: Organizations can only design complex systems that mimic the communications structures of the human organization.

It's also true that in your typical corporation, the top of the management hierarchy is seen as the default landing place for "accountability" (which means blame) when something isn't right. Consequently, middle managers can be some of the least receptive people to the idea that a crisis can be a good thing. To many, "crisis" only means negative attention, and they will go to great lengths to deny that a crisis exists—and, when that fails, to cover it up. This was a factor in many disasters that really got out of hand, including the meltdown at Chernobyl, the loss of the space shuttle *Challenger*, and the recent quality control problems at Boeing.[1]

We'd expect that most middle managers who stumbled into this book have read the whole thing looking for clues about how to "control" a crisis (Translation: "how to make sure nobody finds out about it"). Perhaps a few are realizing that their current job is incompatible with doing real work on the systems they "own," and are considering a change. If this sounds familiar, this chapter may be particularly useful to you.

Or, you may be in the kind of job that is described as "horizontal" or "cross-functional": project manager, chief of staff, compliance, communications, and so on. You're probably already accustomed to jumping from fire to fire as they break out in random organizational hot spots.[iii] People with these kinds of jobs can benefit the most from getting better at crisis engineering.

---

iii Many such people also jump *away* from each fire when it breaks out, but again, by hypothesis, they are not here reading this chapter.

> An interesting paper by Ruthanne Huising in *Organizational Science* followed employees who were displaced from a "central operational role" to a "peripheral change role." Afterward, they tended to see their previous job in the ordinary hierarchy as ineffective. Huising hypothesizes that, rather than people at the outskirts being motivated to change the status quo to improve their own standing, it's that people with the desire to challenge the status quo move themselves to these peripheral social locations.[2] If you are in a long-term project to work your way up inside a large company or agency, this may point you toward the type of role to look for.

FIND A PLACE...

A sufficiently large organization tends to evolve a specialized role or a team that is recognizable as crisis engineering as we have defined it. The original construction of site reliability engineering at Google was just such a thing: a tiny set of people that were empowered to take a wide range of actions without going up and down the chain of command. You may have such a team nearby. You will have to search for it by function; the titles are not standardized and may be anything under the sun.[iv] Of course, all organizations of all sizes have the *function* implemented in some way, or they would have gone extinct. It would not be entirely wrong to say that below a certain size, the *entire organization* consists of crisis engineering. This will make sense to people who have worked in startups.

If you are considering an existing team, beware of the crisis engineering life cycle that we have described. A crisis effort is temporary by nature. Large organizations cannot learn this. They will inevitably see the success that resulted from the temporary suspension of

---

iv We wouldn't be mad if they all came to be called crisis engineering, but the job is amorphous and resists standardization, so they probably won't.

regular order and decide that the thing to do is write down everything that just happened and make it the regular order. The result is that zombie crisis engineering organizations abound.[v]

...OR MAKE YOUR OWN

Last, there's the thing we ourselves are doing.

Through random events, each of your authors was dropped into situations that we now define as a crisis. We learned how to operate in these environments by trial and error. Then we were asked to stay and repeat our magic formulas in many more situations where they did not work. That led us to head out on our own as a consulting firm, looking for those rare opportunities where we could feel like we fixed something. Sorting through the successes and failures motivated us to work harder at defining the conditions where our methods worked. It seemed to us that there was plenty of work for a hundred versions of Layer Aleph to keep busy, if only the system owners knew what we did and why.

That is what led to this book. As far as we are concerned, it is great news if more people want to enter the field. It seems likely to us that our world's need for it will only grow.

HOW MUCH CRISIS IS TOO MUCH?

If you have any success seeking out opportunities for crisis engineering, it won't be long before you are concerned with having too many of them. You will need to take care and prioritize rest and recovery in between projects.

We have argued that to do the most good, it is necessary to match

---

[v] We don't have a good metaphor for this, but a corny one might be the "Sorcerer's Apprentice" segment of *Fantasia*. The magic water-carrying broom is a solution when there is one of them. Then it becomes the problem when a less skilled sorcerer creates a thousand, all mindlessly copying the first one's actions.

the intensity of effort to the level of change tolerance in a complex system. It is not crazy to suggest that if the three of us (plus the others on the repair effort) had prioritized work-life balance and self-care in 2013–2014, there would have been 49 million people in the United States unable to access health care in the last decade, with all the consequences that brings.[3] But it came at significant personal cost. None of us could have managed the physical or emotional workload for much longer than we did.

Clearly, how much is too much varies by individual, and for a given individual, it is different at different points of their career. Our experiences suggest that working in-house, on a team in a large organization, or as a lone specialist in a small one, works well for your first decade or so. You will acquire any missing skills faster when there are more senior practitioners around, and a regular paycheck is a big help when applying for a mortgage.

Mid-career people, let's say years ten to twenty, are in for a hard time if they are part of a big organization and have developed a talent for crisis engineering. Intense, on-again/off-again work schedules are hard to manage when you have a family or other priorities. This is part of the reason why breakout-style organizations expand and cool off. Crisis engineers that successfully stay in one place tend to transform the environment around them, until there isn't as much crisis engineering going on.

We are mid- to late-career ourselves, and we can report that at this point, three or four serious crisis interventions is enough for a year. We limit most of our projects to two weeks. Under the duress of the pandemic, we again found ourselves working for about six weeks at a time, but any longer would have forced us to slow the pace and make time for all that pesky sleeping and eating again.

We've all been learning at our own pace to improve our own capacity and resilience with better mental and physical health. That is out of scope for this book, but good news—that is not only another book, it is a whole other section of the bookstore. We will leave it at this: If you're not taking any of the advice you are hearing along the lines of "take care of yourself," you will find one day that you are forced to, and you will wonder why you didn't start sooner.

Section 4:

CRISIS ENGINEERING IN ACTION

4.1

Ending California's Pandemic Unemployment Backlog

> There was no use putting a little fire out of its misery too soon when you would be paid overtime.
> —NORMAN MACLEAN, *YOUNG MEN AND FIRE*

This chapter walks through a real Layer Aleph crisis engineering example: ending the backlog of unemployment claims in California's Employment Development Department (CA EDD) during the height of the COVID-19 pandemic. We'll show you how we stood up a crisis engineering center, mapped an unfamiliar system, found our people, took novel actions, changed the story, measured progress, and helped the agency emerge stronger and with more capabilities than before.

Normally, all Layer Aleph engagements are strictly confidential—something we take seriously. However, we can share this story because California already made our report public.[i] This chapter is not exhaustive, as that would fill an entire second book.

In 2020, the entire country experienced a massive surge in

---

i This story may sound familiar if you have read *Recoding America* by Jen Pahlka or *Abundance* by Ezra Klein and Derek Thompson. Jen was a cochair of the EDD task force.

unemployment, and therefore in claims for unemployment benefits, as many businesses closed and governments issued stay-at-home orders in an attempt to slow the spread of the COVID-19 virus. Most states struggled to process this surge in claims, but California's struggle was particularly acute: Its claims backlog was on pace to be cleared in forty-eight years. With the process improvements that we were able to make under crisis conditions, the backlog was instead eliminated in three months. We did this while improving the end user experience, not by hiding unprocessed claims in closets. We also used the crisis window to help CA EDD make other transformative changes to its organization.

CRISIS INDICATORS

Let's quickly affirm that this situation counts as a crisis—a window of potential for transformative change—according to our criteria:

- **Fundamental Surprise**—The pandemic was a textbook example of a surprise that overwhelmed everyone's emergency-response plans. Once underway, the surge in unemployment and lag in claims processing was foreseeable. But the size and speed of the wave of new work took the organization by surprise.
- **Failure of Sensemaking**—The claims-processing process broke down. Call center infrastructure couldn't even handle the number of lines needed to keep people on hold. Established performance metrics failed to capture reality.
- **Degradation, Disruption, or Complete Change of Core Processes or Outcomes**—More than 5 million people went months without payments or any idea when they might see a payment.[1]
- **High Visibility**—The backlog was at the top of the news every day, and was the relentless focus of state legislators and the governor's office. Parallel backlogs were growing in other states, but California's was the largest and

therefore the most visible. Some claimants lost their homes while waiting for benefits, and others died by suicide.[2]
- **Rigid Deadline or Timeframe**—The federal government sets a deadline for paying unemployment claims: 87 percent of claims must be paid within twenty-one days. California knew it was now far from hitting this measure, while different factions argued over exactly how far off they were.

A NOTE ON BACKLOGS

Sometimes, as here, a backlog is the root cause of a crisis, and is directly what people are complaining about and want resolved. Other times, backlogs creep up in seemingly unrelated crisis circumstances, catching an organization off guard.

We have successfully unwound some of the largest backlogs in the world, and believe crisis engineering is a repeatable process for measuring and ending backlogs of all types.

What Even *Is* a Backlog?

A backlog is a real *or perceived* processing time below that which is expected.

What Can Be Backlogged?

Almost anything can be backlogged:

- Customer support tickets
- Jira tickets
- Code deployments
- Bug reports
- Benefits claims

- New hires
- Orders
- Payments
- Parts
- Refunds
- Responses
- Policy or training updates
- To-do lists

STANDING UP THE CRISIS ENGINEERING CENTER

Once we arrived on-site the process of standing up the crisis engineering center followed the usual patterns:

- **Convening Authority**—The governor's office declared a "Strike Team" to resolve the backlog.
- **Incident Lead and Team**—The strike team was officially cochaired by Secretary of Government Operations Yolanda Richardson and civic technologist Jen Pahlka. Layer Aleph made up the rest of the initial team.
- **Designated Venue**—We had a conference room in the main agency building. Despite the pandemic, most employees were still working in person, because there was inadequate remote infrastructure to enable remote work.
- **Decision-Making Authority and Permissions**—The day we arrived, we had badges that let us into the building and user accounts with access to most machines (we made additional access requests over the course of the engagement, which were quickly granted). Secretary

Richardson provided decision-making authority as needed.
- **Low-Latency Communications**—Impromptu meetings took place throughout the day, with daily coordination standups between Layer Aleph and the other parts of the strike team.
- **Shared Journal**—We kept a running log of actions taken in a shared file.
- **Announcement and Kick-Off Ritual:** The governor's office issued a press release, introduced the strike team to the legislature and media, and encouraged all employees to talk with us.

MAPPING THE SYSTEM

We started the process the same way we usually do: by mapping the process for submitting and processing an unemployment claim. The goal was to converge on one version of the truth across stakeholders who, we found, had very different views of reality.

On the way to our assigned conference room, we noticed huge printouts of the claims process, represented by perfect squares and arrows. We also saw a poster announcing "innovative" new changes, decorated with interlocked gears that, if they were real, would be immovable. This is such a common trope that we maintain a photo album of posters from our projects featuring gears that could not possibly turn.[ii]

Confident no one was using these posters, we took one of them down and started drawing our own map on the back.

Usually, two of us walk around the building(s) where the work takes place. When needed, we drive around a metro area or fly around the country to get to a physical place where we can see the

---

ii Let it be noted that the gears on the cover of Marina's book *Hack Your Bureaucracy* could, in fact, turn.

work up close. That's what we did here, too, with the added challenges of in-person work in the height of a pandemic (before vaccines existed), a deadly heat wave, and an active wildfire burning just outside. On our second day, a wild turkey tried to prevent Marina from entering the building.

We sought to understand the process from the applicants' view as well as from the operators'. Once we had one end-to-end path, we went back to the start, looking for variations and asking people to give us mysteries. For example, upon learning Marina was the former CTO of the VA, one claims processor recalled a mystery: Why couldn't veterans file online?

We asked around and found a claim from a veteran to follow. The nuance did not end up being whether someone was a "veteran" but rather if they had just separated from the military. In this case, their proof of recent employment is a Department of Defense form called a DD-214. The website did not allow document upload. Since veterans had to attach their DD-214 form, they could apply only via mail. In the pandemic backlog, mailing in an application was approximately as effective as submitting it via smoke signal.

Another critical discovery from our mapping process involved the call center. In our interviews, almost everyone told us about the call center and how a claimant with a problem could resolve it by calling in and speaking with an experienced human. The agency promoted the phone number to claimants. The media shared it.

We drove to the call center...only to find an empty cubicle farm, with one lone manager sitting in a side office. He was very friendly and happy to answer our questions, but was very confused as to why we were there to talk about the call center. This was not the call center, he assured us. He knew about the call center, and gave people its number regularly, but he was sure it existed somewhere outside his office.

What happened? The whole time, the "call center" had simply been a number that connected to a random employee's desk. Prepandemic, employees answered their phones and tried to answer callers' questions. They did not think much about how the caller got their number, and they did not realize they were a "call center." When these employees began working from home, the phones kept ringing, but nobody answered them. And nobody noticed, because no one ever had looked for the other end of the hotline. In 2013, there were HealthCare.gov call centers in similar nonfunctional states. This is why we are so insistent on you physically going and seeing every step for yourself.

Zooming In and Out on Identity Verification

As we mapped, we found some helpful automation. But once a claim went off the automation track, it could never get back on—every subsequent step had to be completed by a human.[iii] Humans were in very short supply, and extremely difficult to scale: Many tasks required at least sixteen years of experience (because of accounting credentialing levels).

The available workforce could never work down the backlog manually; automation was the only way out. Failing identity verification was the top reason for falling off the automation track. This led us to zoom in on this subsystem.

The new pandemic benefits programs no longer required the standard process of verifying your past employment. Being a real person who lived in California was almost the only requirement to qualify. (This wasn't a problem—it was on purpose.) That shifted all the responsibility for fraud prevention onto the identity verification team.

The automated process approved you as a real person if you provided a name, Social Security number, and other identifying information that *exactly* matched the master list in the mainframe. If there was any discrepancy, such as variations on first names (Kathy versus Kathryn), it was routed to the human identity verification team.

---

iii Most processes don't reintroduce automation once humans have gotten involved, per the Law of Requisite Variety we explained earlier.

If you failed the automated identity verification, you had to mail in a copy of your driver's license as proof of identity. According to the CA EDD's PowerPoint slides, the identity verification team carefully scrutinized these documents to prevent fraud. No one wants fraud, so this function had run unquestioned for a long time.

At face value, this plan isn't crazy. The trouble is that the implementation didn't remotely resemble these simplistic descriptions.

The mailed copies of driver's licenses were scanned by a vendor and saved to a separate system for the identity verification team. The vendor contract specified the exact colors that would be scanned. This list did not include the color red. Unfortunately, both ID number and birth date are red on California licenses. This meant the team literally could not see two of the most important pieces of data—if they could read anything at all from a scan of the print of the copy of the license.

The team was left to guess whether "John Smith" or "Vy Ha" were real people,[iv] and other such problematic questions. When asked how they verified out-of-state drivers' licenses, the team confidently told us: "With our eyes." If an address was associated with more than a few applicants, they flagged everyone as fraud; sadly, when we looked up these addresses individually, they were mostly associated with homeless shelters.

When we asked to see examples of fraudulent documents, they pointed us to stacks of returned mail. Undeliverable mail was considered evidence of fraud. Upon inspection, every envelope in this pile seemed to be missing the second half of the mailing address. Further investigation found that the database field for "street address" cut off at thirty characters, so practically anyone living in an apartment building with a long street name never got their mail.

Zoomed in, team members saw themselves as the last line of defense against fraud. Lacking clear or practical tools, and not

---

iv Patrick McKenzie wrote a list in 2010 called "Falsehoods Programmers Believe About Names." Without getting too specific, let us say that civil servants also believe these same falsehoods about names. (https://www.kalzumeus.com/2010/06/17/falsehoods-programmers-believe-about-names/)

having any actual expertise on identity verification, they tried their best, with a bias toward "better safe than sorry." This meant rejecting a large number of claims because from their perspective, "all" the claimant had to do to try again was submit more documentation. They interpreted the low resubmission rate as evidence that they were actually stopping tons of fraud.

Zoomed out, however, identity verification was taking down the whole ship.

While the human team introduced tremendous bias and delay—enough to fundamentally break operations on its own—the automatic acceptance of applications that precisely matched the mainframe was causing even more harm. Lists of names and Social Security numbers, stolen directly from government mainframes, are easy for criminals to get. Organized crime groups, mostly outside the United States, assembled their own automation to submit mass numbers of fake applications that precisely matched the government data. The U.S. Department of Labor eventually estimated the "overpayment rate" at 17 percent, or around $22 billion.[v] Opponents of unemployment benefits began using this actual fraud as grounds to propose the complete elimination of the program.

Zooming out revealed that identity verification had actually *never* worked. Before the pandemic-specific unemployment program, the agency had to directly contact and confirm your employment with your previous employer. *This* step was the only one that ever verified you're the person you claimed to be, and was what was preventing fraud at scale—because it's hard to fake tens of thousands of employers who answer the phone. With employer verification suspended, the remaining steps immediately failed, exposing the entire organization to a tsunami it was not able to handle.

---

[v] United States Department of Labor, "PUA Improper Rate Report," August 21, 2023. "Overpayment rate" is a bit of bureaucratese. Everyone knows it is mostly fraud. But there is a little bit of "honest mistake" overpayment, too, which enables the agency to insist that 17 percent is *not* the rate of fraud.

FINDING OUR PEOPLE

As we mapped the system, we found many people who shared details, expertise, and insights for their part of the system. We also found many points of resilience. For example, the two-person team in charge of the automation scripts were constantly talking with their business counterparts to come up with new scenarios that could qualify for automated processing. This collaboration, which literally kept the entire system from collapsing, was never formal or documented. These two developers could not even get their request for a couple of extra computers approved.

We mentioned in chapter 2.4 (Find Your People) that Marina's purse often serves as a sort of mailbox for reaching us. This happened almost daily on this project. For example, we sat in one meeting where a manager firmly told us there was no way to know how long it took a claims processor to work particular steps of the process. She couldn't even give us a back-of-the-napkin estimate. In truth, they published weekly performance reports, broken down to the minute by operator—a clue Marina found in her purse an hour later. We also received screenshots of email inboxes with ten thousand-plus new messages (mostly from new hires with questions or requests to be unassigned from a claim they couldn't handle), handwritten pleas for help, and invites to stop by certain desks.

TAKING NOVEL ACTIONS

CA EDD was entering the tenth year of its Benefit Systems Modernization project, which had pulled all the best operators away from processing unemployment claims, in order to list requirements for a perfect future system. Like all modernization projects, it was behind schedule and nowhere near even *starting*, yet even the tiniest of changes to the current system had been indefinitely put on hold until the launch of this new system.

While formally on hold, *some* action was definitely still happening.

The two guys mentioned above were changing code right and left to accommodate the new pandemic-specific rules. We had to persuade the agency to pause its modernization project and lean into bold actions.

Some of the actions we prompted:

- **Changing the Identity Verification Process.** Both identity verification issues covered earlier—an automated flood of fraud, with real people caught in a tar pit of ineffective manual practices—were addressed by outsourcing identity verification to a vendor (ID.me). Now, real claimants could verify their eligibility automatically using industry-leading practices and technology (not eyeballs). The vendor's process included "escape hatches" so that real humans could still access other humans to resolve edge cases like expired driver's licenses or limited English proficiency.
    - We don't advise clients to take on new vendor dependencies lightly; they chronically underestimate the real costs and overestimate the benefits. But in this case, it seemed like a net positive to ensure that the agency employees were unable to touch the identity verification process in any way. If they retained access, we worried they would find ways to reinsert the faulty heuristics they believed in.
    - We carefully watched the dashboards after launch; it took some explaining as to why the automation rate went *down* (because all the fraud that was being automatically processed before was no longer going through). But over days, all signs pointed to increased automation of nonfraudulent claims. This automation was one of the two big levers for ending the backlog (the other being the self-service status tracker).
- **Turning It Off.** Our analysis accomplished the unthinkable in California: It persuaded leadership to stop accepting new applications for two weeks. The

public wasn't exactly happy about this, but the pause created breathing room for the agency to catch up, implement the change to outsourced identity verification, and stop adding to the pile of avoidable extra work like duplicate submissions and support requests. Our data model, that showed anyone who applied after the launch of ID.me would be processed *faster* than if they applied two weeks earlier in the original process, was persuasive.

- **Moving People Around.** In a great example of why sensemaking requires action, we were surprised to see that the day after reassigning many employees to Step 1 in the claims process (a step that required only minimal training and around six minutes to complete), the claims automation rate plummeted to near zero. Recall that in this system, a claim could never return to the automation pathway once it became manual, so this had huge downstream repercussions on the volume of work requiring already-stretched-thin humans.
    - Everyone assumed a bug in the automation code, but when we looked at that code, we learned an important new fact: The automation looked for new claims only every thirty minutes. The new employees were grabbing claims for the manual process *faster* than the automation could pick them up. Oops!
    - Changing the script so humans could only access new claims *after* the automation script rejected them solved this issue, and automation rates went back up.
- **Expanding People's Scope of Action.** Identity review was just one of many tasks that went to human operators. Hundreds of other triggers could cause an application to stop and wait for a human. These were tracked in a system that maintained a separate "work queue" for each type of task. Rules for accessing and moving items from work queues were very, very specific. Any given

employee was permitted to take items from only certain work queues, determined by their level of seniority, pay grade, job title, and place in the org chart. Basic queueing theory tells us that such differentiation can only slow things down.

- Some of the queue separation was desirable; it's true that you may not want people who started last week handling advanced scenarios. But much of it was unnecessary. For example, employees who had handled a certain work queue in a past assignment could do so again without any new risk. We proposed reinterpreting the work queue rules to allow caseworkers to practice to the full extent of their "license," rather than be as restrictive as possible.
- This change felt like rolling downhill because all we did was remove obstacles to things people *already wanted to do*. Much of the time, changes that improve the distribution of tasks among humans and machines will be like this.

- **Rewriting in Plain Language.** The fact that so many people submitted multiple claims signaled to us there was a process challenge beyond the backlog itself: Applicants were not getting a confirmation message, not hearing back soon enough, getting confused, or trying to update an existing claim with new information. All these duplicates were making the backlog worse. Upon review, there were many opportunities to clarify instructions. Eliminating unnecessary duplicates helped reduce system load.

WHY HIRING THOUSANDS OF NEW PEOPLE PARALYZED PRODUCTIVITY

A classic example of well-intentioned but counterproductive action in a crisis is rapidly scaling up the workforce. Before we arrived on the scene, California had already hired thousands of new employees to help work down the backlog.

The thinking was simply that more hands equaled more work done. In reality, the hiring rush overwhelmed the organization's already-strained onboarding and training processes. Managers who should have been focused on fixing systemic issues were diverted to supervise and train the newcomers. Knowledge transfer from experienced employees was chaotic and incomplete. The result? Productivity plummeted further—in some cases to literally zero—and bottlenecks worsened.

The lesson here is not that hiring is always wrong, but that it must be targeted and aligned with the organization's capacity to absorb and utilize new resources effectively. Reverting a one-line code change is (hopefully) quick if you change your mind; unhiring one hundred people is not. Scaling too quickly without addressing the underlying systemic issues—or without a plan for integrating and empowering new hires—can exacerbate dysfunction.

While we have occasionally recommended a handful of strategic, urgent hires (usually, bringing an experienced retiree or former contractor back to the scene), we have never seen a mass hire make a crisis better.

MANAGING THE STORY

At the start of our engagement, there was a count of backlogged claims, and a goal: 87 percent of claimants who were entitled to payment must receive that payment within twenty-one days. People were furious and at the end of their ropes, and success seemed impossible.

At the end of our engagement, there were still backlogged claims, and the goal had not changed. But crisis conditions were receding. Tempers cooled down, and success seemed assured. This was validated a few months later when the agency successfully worked down the backlog.

How did we change the story while the backlog still existed? One thing that helped was to evolve the definition of "backlog" so that it made more sense to the operators responsible for working it down, as we discuss in the next section. That set up for the real turning point, which was when the backlog number plateaued and began to come back down. People's emotions are much more responsive to the direction this sort of metric is going, rather than its absolute value.

MEASURING PROGRESS

Measuring the Backlog

Per federal requirements, 87 percent of claimants who were entitled to payment must receive that payment within twenty-one days.

But that left a lot of flexibility for which claims could count as "backlogged." The news and legislature believed there was a huge backlog, but CA EDD worked hard to argue that certain categories of claimants shouldn't factor in.

For example, it was a requirement that claimants reconfirm their unemployment every week. If someone did not reconfirm in the last week, the agency stopped counting them in the backlog. Given the pandemic, it seemed unlikely that all these claimants were suddenly finding employment and no longer in need of benefits. Rather, they were probably confused, frustrated, or, as it turns out, actually reconfirming all along—the agency was behind on opening the mail. These claims should *always* have counted in the backlog total.

You can argue that your self-defined backlog is as small as you want, but if the legislature and the news media are deluged with complaints, nobody will believe you. In this case, it led to escalating pressure, in the form of angry news articles and legislative inquiries. The legislature lost so much trust in the official number that it began tracking its *own* list of backlogged claimants, which grew so large that it became another IT system we had to integrate into the backlog tracker we built.

By the end of our engagement, we had an automated, consistent, plausible count of backlogged claims—a *shared story*. Here's how we did it:

- **We Started with One Big Bucket.** We did not waste any time talking ourselves into why this or that pile didn't count. We looked across paper files, the mail room, faxes, emails, handwritten steno pads on desks, support tickets, and legislative complaints in addition to the multiple IT systems, including a system log that doubled as a work queue.
- **We Worked with All Corners of the Business to Decide What Counts.** This was an iterative process. Participants were inclined to come up with creative reasons why certain applications didn't count as backlogged: if the agency was waiting on them to send something back in, or if the timeframe dictated in policy technically expired (even though with the backlog, the agency wasn't holding up its side of the bargain on these timeframes).
  ○ Each time the agency wanted to exclude a category from the backlog, we asked the question from the

other side: Does this feel like a "backlogged" claim to the applicant?

- **We Began Instrumenting the Count.** This simply means bringing all the measurements together and combining them. At the start, this was a huge mess of spreadsheets and database exports. Mikey had to pull data from over seven systems and work with many claims processors to understand what more than a hundred specific codes meant in each area. But having numbers on a screen put stakeholders in the position of having to explain why *this* number was wrong, specifically, rather than just rail against the concept of numbers. This shook out lots of details that made the count more accurate.
    - One of the last holdouts was the mail room. Groupthink held that there was no way to count claims or track ID numbers in the bins full of unopened mail. There was a period where no mail was being opened at all, because all those workers had been directed to work from home. This evolved into a moral hazard: Whether conscious or not, caseworkers had an incentive to bounce claims into the mail pile, where they would never be counted again. Mikey found a way to measure unopened mail: a scale. He personally weighed all the bins and added "pounds of unopened mail" to the backlog dashboard.
- **We Refined the Count.** When we started pulling together the claims information across all the different systems, we knew there were tons of duplicates, among other challenges. We simply wanted to get our arms around the universe of claims. Now we wanted to find unique identifiers across the information so one claimant counted only one time. In this way, the original backlog count had been unfairly inflated, because many

desperate people had applied tens of times each. This often happens with backlogs.
  - The agency had nothing to lose by redefining the count during the crisis, because nobody trusted it anyway. Any other time, they would have been dragged to the legislature to explain any dramatic differences. We encouraged them to use this window to get to the most accurate count possible, since they'd be stuck with it for the long term.
- **We Automated the Count.** Handcrafted numbers are dangerous and vulnerable to manipulation. Once we got through all the manual data merging, deduplicating, and related tasks, we worked with an existing data team to automate the process. This ultimately generated a backlog count every night on its own, without human involvement. (Funnily enough, the "pounds of unopened mail" put so much scrutiny on a previously ignored part of the process that all unopened mail was quickly handled, and this final manual measurement disappeared.[vi])

Making a Burndown Chart

We had one, accurate, agreed-upon story of how the unemployment claims process worked and how many people were in the backlog. Now, we wanted to rapidly model the impact of different actions we could take to *change* that story.

When we started this gig, the agency could not estimate how long it would take them to work through the mountain of claims. We worked with them to develop a burndown chart (sometimes called a stock and flow): a rough model of the system (made in a spreadsheet) showing how work flows through the system at each step. This is different from the backlog count, as instead of considering how many

---

[vi] A good example of the exploitability of Goodhart's Law from chapter 2.7 (Measure Progress).

people were negatively impacted, this considered the total amount of work ahead.

The first chart showed it was going to take *forty-eight years* to work down the current claim volume. When we left the project a few weeks later, the estimate had dropped to three months. Three months later, we got a message from the team that they had ended that backlog!

Here's how we built the spreadsheet:

- **Listed Each Step of the Process.** Steps included an operator "sweeping" a new application into the system, the mail room opening mail, mail being scanned, identity verification, etc. We listed all major steps, even if they didn't apply to every claim. One step was roughly "what one operator does to one claim at one time."
- **List How Many People (and Hours) Are Assigned at Each Step.** The time study report someone had slipped in Marina's purse provided the starting point of how many people were assigned to a step. We used a one-week timeframe.
  - The report helped us avoid the common mistake of assuming one full-time employee works forty hours a week; that's not true when you consider breaks, lunch, checking email, and work-related tasks that are not directly related to a given processing step.
  - We were particularly careful *not* to count any of the new hires, who could effectively do no work. We first used this burndown chart to model how all those new people were taking time from experienced people, leaving zero processing power at some steps.
- **How Long Does It Realistically Take to Complete This Task?** The above report also indicated how many minutes each step took, so we started there.
  - For steps where we were unsure of the time necessary, we went back and asked the people who

performed that role to give us their best guess. We knew we could improve these estimates over time as we watched how the model performed.
- **What Is the Volume at Each Step Right Now?** Mikey was generally able to pull these numbers from the different systems in the course of generating the backlog reports. He also confirmed the numbers aligned as claims moved between steps; we often find a "seam" when working on backlogs, where some number of items from Step 1 literally never make it to Step 2, and no one notices the discrepancy.
- **Was There Lag Time Between Steps?** For example, a batch job ran nightly at 1 a.m. The agency could never process faster than the batch job runs (or, the batch job could run more often).

Additional considerations for our chart included:

- **Subsequent Tasks.** Each step in the process generated additional steps downstream. For example, if an application went to manual identity verification and was approved, it would then create a series of additional processing steps. The chart had to model this in order to estimate the entire volume of work. In some cases, we relied on operators' best guesses ("this happens about half the time").
- **Filtering Out Unimportant/Unnecessary Tasks.** Some work items were "nothing burgers"—an informational or a useless alert employees simply had to click past. We filtered these out, as it was not work that actually needed to get done. (Separately, we helped identify ways for a script to clear these out automatically.)
- **Tasks That Increased or Decreased Based on Processing Speed.** The slower the agency was to get back to someone, the more likely they were to resubmit their

claim, creating more work. As the agency sped up, we expected these duplicates to disappear.

We shared the burndown chart (and often the system map) with leadership and other members of the organization, both to validate its current setup and to generate ideas for ways to change it. After all, forty-eight years was not an acceptable timeframe.

We encourage taking actual actions, but once we had this chart, we would quickly model the impact of our intended change.[vii] At first, this mostly revealed ways in which the burndown chart contained incorrect assumptions. But as it grew more reliable, it became a valuable tool for persuading leaders they had to take drastic, immediate actions (like completely replacing the identity verification process, or shutting down the call center).

The burndown chart had an additional advantage: After the crisis, we were told they continued to use it to refine workforce allocation and keep an eye out for early issues in any individual part of the process.

COMMUNICATING IN THE CRISIS

The internal communications on this project were pretty standard, but the biggest opportunity for transformation and improvement here was in how the agency communicated with the public (for example, people who filed for unemployment).

Ninety-nine percent of calls to support were inquiring about claim status.[viii] Unfortunately, the support agents had no information beyond the claimant's status code, which was something like "F34." This was not helpful; it positively *enraged* people who had sat on the phone for hours waiting for help.

---

vii A pet peeve of ours to keep in mind: Saving six minutes across twenty thousand employees gives you back zero minutes. If a step takes two hours on average to complete, and you knock off a few minutes elsewhere, we wouldn't expect you to get even one additional work item of productivity. This does not mean the automation of a six-minute task isn't still worth doing!

viii That is the actual number. The remaining 1 percent was for password resets.

After enhancing identity verification, it became clear the next big immediate action was to provide a self-service claim status tracker. Critically, any message needed to translate a status code into plain English (and Spanish).

Over a two-week period, the agency set up a digital claim status tracker.

Rather than wrangle with the IT security implications of creating user accounts for this claims status application, or trying to integrate with the existing legacy login system, the application used a simple "name and last four of Social Security number" framework for logging in. A sentence explaining a status code was deemed as not personal or private information, so the team could sidestep extensive security approvals. This is a great example of taking action with existing tools and approvals.

The plain language meant that in cases where the agency was waiting on claimants, those claimants now knew what to do, which moved stale claims along. Converting status codes into easy-to-understand sentences was the hardest part of this operation, as there was no single person who could translate all the codes. Many operators from across the agency had to collaborate and share their pieces of the puzzle.

The tracker leveraged an existing nightly spreadsheet export from the mainframe, so there was no new strain on that machine. (Not only would the mainframe have fallen over and died immediately if millions of Californians tried to check their claim status directly, but the behavior of desperate people is to hit "refresh" over and over, which would have magnified the load considerably.)

Once launched, calls to the support hotline plummeted as people flooded to the self-service tool.

INSTILLING TRANSFORMATIVE CHANGE IN THE WAKE OF THE CRISIS

CA EDD exited the crisis with a new, automated process that could accommodate surges without having to add more humans, which is important in a system that requires its humans to have sixteen years of experience. The agency did not simply adjust definitions, ignore piles of mail, or brute-force its way through. That's pretty transformative for a government agency over the course of a few weeks.

At the close of the crisis engineering engagement, the state canceled the Benefit Systems Modernization project (the one that had yet to deliver anything in ten years) and expressed a preference for a more agile, iterative approach to modernizing its systems.

Getting a handle on fraud saved money directly, and might have stabilized political support for the unemployment insurance system overall.

All this happened through sensemaking:

It's Motivated by a Sense of One's Own Identity. The identity verification team saw themselves as the last line of defense against fraud. This drove a lot of behaviors that seemed rational to individuals, but worked against the delivery of benefits and also enabled fraud. When this was understood, the organization began to think of itself as responsible for administering benefits fairly. Though it may seem similar, this mission statement is radically different from "prevent fraud."

It's Retrospective (Backwards-Looking). Looking back at contributing factors to problems allowed EDD to understand past failures. Mapping the system highlighted that a significant portion of claims were unnecessarily routed for manual processing and subject to outdated identity verification methods.

It's an Active Process of Cocreating Your Environment. EDD actively reshaped its operational environment. They developed internal dashboards to monitor claim processing in real time, allowing for dynamic resource allocation, and a now-shared map helped different parts of the organization see how they fit into the bigger picture in ways they had not before. Actions that drove outcomes in the wrong direction (like hiring thousands of new call center workers overnight) still helped improve understanding of how the agency's processes actually worked, leading to better choices in subsequent actions.

It's Social. Transformation required coordinated efforts across various teams within EDD and external partners. Briefings, shared dashboards, collaborative whiteboard sessions, and lots of conversations fostered a collective understanding and willingness to change. It also required being in person—had everyone tried to solve this remotely, we do not believe this ever would have been solved.

It's Continuous, with No Start or End. Recognizing that transformation is an ongoing process, EDD continuously took new actions and made more improvements. New tools like the backlog dashboard and burndown chart helped ensure adaptability to evolving challenges.

It's Built on Facts That We Select or Deduce from the Environment. The process of mapping the system, physically walking to every step of the process, and automatically calculating the dashboard figures selected the necessary facts to build a new reality.

It's Driven by Plausibility More Than Accuracy. Not every new change or action was perfect—but it didn't have to be. What mattered was that the new claims-processing model felt legitimate, credible, and workable to the people using it. Early wins around identity verification and claim status helped reinforce the plausibility of the new direction, even as the finer points continued to evolve.

In the end, California's unemployment backlog wasn't resolved by waiting it out, doubling down on the status quo, or an endless expansion of staff—it was resolved through disciplined crisis engineering and relentless sensemaking. By mapping reality as it was (not as it was assumed to be), taking nonintuitive but high-leverage actions, and cocreating a shared understanding across the organization, CA EDD turned a forty-eight-year problem into a three-month success story. The same pressures that could have crushed the agency instead forged a stronger, more adaptive system—proof that when you harness the heat of a crisis, rather than run from it, you can build something far more resilient than what came before.

Conclusion

Crises are unavoidable. They are a natural part of the life cycle of a complex system. Organizations in all domains face moments when their environment shifts faster than their systems can adapt. You cannot prevent every crisis, but you can choose how to respond. You choose whether the organization emerges stronger and more prepared for the next adaptation, or merely accumulates more scar tissue.

Crisis engineering can help you meet these moments with clear eyes and reliable tools. It's not about working harder or reacting faster. It's about understanding that every crisis opens a brief window for real change. In these moments, information flows more freely, decisions are made more quickly, and resistance to change softens. The assumptions that have governed an organization for decades can be surfaced, challenged, and replaced in minutes. With the right approach, you can move beyond just surviving the crisis. You can build anew, stronger and more resilient than before.

Throughout this book, we showed you how to recognize the indicators of a crisis and how to harness the unique dynamics at play. We've given you a toolkit to rapidly generate sensemaking, to map reality, to track real progress, and to choose novel actions. We've outlined how to build a crisis engineering center, how to find the right people and ask the right questions, and how to keep acting, learning, and adapting even when the pressure is at its highest.

Most important, we have shown that a crisis is not simply something to endure. It is a rare, invaluable opportunity to engineer change that is impossible under normal conditions. A crisis allows you to question assumptions, reshape what was rigid, and leapfrog past years of slow, incremental improvement. When approached with this intention in mind, a crisis can be the inflection point where an organization stops treading water and starts mastering its environment.

The tools and frameworks we've shared are powerful, but they are only as good as your willingness to use them. When the next crisis comes—and it will—you will face fear, confusion, resistance, and the temptation to cling to the familiar. Fight those instincts. Remember: The destabilization you are experiencing is not just a threat; it is your opening. If you act with courage and clarity, you can bring about outcomes that will reverberate far beyond the crisis itself.

We wrote this book because the world needs more crisis engineers. Our systems are being tested harder and faster than ever before. The ability to make sense of chaos, to act wisely under pressure, and to build lasting improvements in the wake of disruption will define the organizations and leaders that succeed in the future, across social, political, or legal crises as well as technical ones.

You now have the tools. You now know the path. The next crisis will not wait for you to feel ready. It will simply arrive—and when it does, you will be ready to meet it.

Acknowledgments

We are first and foremost grateful to our editor, Natalie Bautista. Her guidance, feedback, and patience have been instrumental to this book becoming something we are proud of. We are also forever appreciative of our agent, Bridget Matzie, whose magic and faith transformed a brief email from Marina into this book you are reading.

We owe debts to our business partner, Carla Geisser, for both getting the ball rolling years ago with her suggestion to create a workshop, and providing manuscript feedback since the days when we had little but a pile of academic papers and an outline.

We would not have done any of this were it not for Anil Dash, who told us we were better than anyone in the world at what we do, but that it had no name and that we should give it one. We worked hard at that, but none of our other ideas outcompeted his original suggestion of "crisis engineering." We're grateful for his impetus, and his review of an early manuscript. Laura Thomson is who told us we should "write the book on it."

What we hope to have communicated in this book draws on experience and research in fields as wide-ranging as engineering, sociology, ecology, anthropology, spirituality, and public policy. We have been particularly influenced by the work of: Melody Beattie, Dan Davies, Anthony Downs, Joseph Henrich, Frank Herbert, Daniel Kahneman, Aldo Leopold, Nancy Leveson, Norman Maclean, Donella Meadows, Peter Sandman, Arthur Squires, Diane Vaughan, Peter Watts, Karl Weick, and James Q. Wilson.

We are humbled by the incredible number of people who gave

us their time and energy to provide ideas, feedback, encouragement, and to review various drafts. We're sure we will have missed someone, but doing our best: Shawn Arnwine, Valerie Aurora, Scott Blackburn, Silvia Botros, Deb Chachra, Camille Fournier, Jason Fraser, Alex Gaynor, Greg Gershman, Dave Guarino, Kavi Harshawat, Dan Hon, Jeff Maher, General Stanley A. McChrystal, Denis McDonough, Tara McGuinness, Patrick McKenzie, Giuseppe Morgana, Tim O'Reilly, Jen Pahlka, Ryan Panchadsaram, Amy Paris, Tanya Reilly, Melanie Rice, Dr. Peter Sandman, Nick Sinai, Paul Smith, Tamara Srzentic, Bob Sutton, Paul Tagliamonte, Emily Tavoulareas, Natalia Villalobos, Seth Wainer, Emily Wright-Moore, and Mike Wilkening.

Marina would like to acknowledge the steadfast love and support of her husband, Charles, who promises he believed her when she said she would not write another book after *Hack Your Bureaucracy* and promises he believes her when she says this again now. I love you and how well you take care of me.

Matthew Weaver is grateful for the steadfast love of a power beyond his understanding, along with the love and support of his wife. He is glad he can look to the example of patience, persistence, and humility provided by his neighbors and fellow travelers.

Appendix A

Layer Aleph's Standard Engagement Model

Clients hire us to make sense of all sorts of crises and systemic problems. While the specific goals and outcomes of each engagement vary widely, the majority of our projects follow a set pattern the first few days.

DAY 1

We start the day with a kick-off meeting. This is an opportunity for the client to set goals and expectations, do brief introductions, and get an overview of "the problem" as the client sees it. It is also when the client signals to the entire organization that something new and unusual is happening, that there is a time-limited suspension of normal order. They tell everyone to make themselves and their information available to us as an overriding priority.

We then move to brief, informal presentations by the people closest to the problem describing how things currently work. These may include past risk assessments, architecture diagrams, and process walkthroughs. We insist that clients not create new presentations just for us. We're happiest with a grab bag of content repurposed from existing work and unfiltered takes from experts.

DAY 2 AND ONWARD

Starting with an outline of the organization's complex system, and team members from Day 1, we'll ask questions, read documentation, and generate ideas. Our goal is to iterate and make hands-on changes in partnership with the client's team while expanding our map of the system.

We will join any regular meetings where a client asks for our participation and engage with new incidents as they occur during the engagement's time window.

We aim to be spending most of our time with people who are hands-on experts by the middle of the first week. These people can be claims experts, customer support agents, database administrators, or software engineers who are keeping the client's systems and processes running. They are the source of our best stories and insights. We'll ask the people we meet for introductions to others as we work our way through the client's organization.

DAILY CADENCE

Most clients like to set up a daily morning and evening check-in with us. These are typically fifteen to thirty minutes with the principals to share new questions, early findings, and roadblocks.

THE END

We'll use the last day of our intensive engagement to present our recommendations to leadership and key team members.

NOTES ON OUR STRUCTURE

Layer Aleph is small, and there is no hierarchy among the partners. We each have particular areas of expertise and interest that we'll match to the client's concerns during the engagement. We'll typically have one person act as the primary client contact during a project. We are in constant communication with one another to avoid duplication and share insights, even when we split onto different tracks of work during the day.

PREREQUISITES

We are most effective when we can access client systems directly, ideally by the end of our first day of work. The most important tools for us to be able to access are:

- Real-time communications (Slack, Microsoft Teams, IRC)
- Document repositories (SharePoint, Google Drive, Dropbox)
- Monitoring consoles (New Relic, Grafana, Google Analytics)
- Source code (GitHub, GitLab)
- Task and issue-tracking systems (Jira, Microsoft Project, ServiceNow, GitHub)

DELIVERABLES

We negotiate expected deliverables with clients ahead of time and put the details in our signed statement of work. We prioritize actionable recommendations and will make changes (with permission) directly to client systems to drive improvements during the engagement. We find brief, highly actionable memos to be more useful than long reports, but if a client requires a longer written assessment, process map, or architecture diagram, we can absolutely provide one.

ADMINISTRATIVE SUPPORT

When working with particularly large organizations (thousands of employees) or wide surface areas for assessment (multiple C-suite stakeholders), we may ask for help from an executive administrative assistant or project manager. These people are extremely helpful for understanding the hidden organizational structure and getting time with key experts.

Appendix B

Recommended Reading and Bonus Resources

For a crisis readiness assessment, recommended reading list, and more resources, please visit our website at CrisisEngineering.com.

Notes

0.2: What Is a Crisis?

1. M. W. Seeger, T. L. Sellnow, and R. R. Ulmer, "Communication, Organization and Crisis," *Communication Yearbook* 21 (1998): 231–275.

2. Zvi Lanir, "Fundamental Surprises," Center for Strategic Studies, University of Tel Aviv (1984).

3. Karl E. Weick, Kathleen M. Sutcliffe, David Obstfeld, "Organizing and the Process of Sensemaking," *Organization Science* 16, no. 4 (2005): 409–421, https://doi.org/10.1287/orsc.1050.0133.

4. Karl E. Weick, "Cosmos vs. Chaos: Sense and Nonsense in Electronic Contexts," *Organizational Dynamics* 14, no. 2 (1985): 51–64, https://doi.org/10.1016/0090-2616(85)90036-1.

5. Norman Maclean, *Young Men and Fire* (University of Chicago Press, 1992).

6. Anthony Downs, *Inside Bureaucracy* (Little, Brown, 1967), chap. 13.

7. Steven James Venette, "Risk Communication in a High Reliability Organization: APHIS PPQ's Inclusion of Risk in Decision Making" (PhD diss., North Dakota State University, 2003), https://www.scribd.com/document/889774778/Risk-Communication-in-a-High-reliability-Organization-APHIS-PPQ-s-Inclusion-of-Risk-in-Decision-Making.

0.3: What Is a Complex System?

1. Christine Sinsky et al., "Allocation of Physician Time in Ambulatory Practice: A Time and Motion Study in 4 Specialties," *Annals of Internal Medicine* 165, no. 11 (2016): 753–760, https://doi.org/10.7326/m16-0961.

2. W. Ross Ashby, *An Introduction to Cybernetics* (London: Chapman & Hall, 1956).

1.1: Sensemaking 101

1. Harold Garfinkel, *Studies in Ethnomethodology* (Prentice Hall, 1967).

2. D. J. Devine, L. D. Clayton, B. B. Dunford, R. Seying, and J. Pryce, "Jury Decision Making: 45 Years of Empirical Research on Deliberating

Groups," *Psychology, Public Policy, and Law* 7, no. 3 (2001): 622–727, https://doi.org/10.1037/1076-8971.7.3.622.

3   Shai Danziger, Jonathan Levav, and Liora Avnaim-Pesso, "Extraneous Factors in Judicial Decisions," *Proceedings of the National Academy of Sciences* 108, no. 17 (2011): 6889–6892, https://doi.org/10.1073/pnas.1018033108. As usual, the conclusions are disputed. But no one seems to argue the effect doesn't exist; they only argue about its magnitude and cause.

4   Jessie Singer, *There Are No Accidents: The Deadly Rise of Injury and Disaster—Who Profits and Who Pays the Price* (Simon & Schuster, 2023), 91, Kindle.

5   James Q. Wilson, *Bureaucracy: What Government Agencies Do and Why They Do It* (Basic Books Classics, 1991), 62.

1.2: Three Mile Island

1   Mike Derivan, "TMI Operators Did What They Were Trained to Do," *ANS Nuclear Cafe*, April 23, 2014, https://www.ans.org/news/article-1556/tmi-operators-did-what-they-were-trained-to-do/.

2   United States. President's Commission on the Accident at Three Mile Island. Public's Right to Information Task Force. *Report of the Public's Right to Information Task Force* (Washington, DC: Nuclear Regulatory Commission, n.d. [1979]), 90, PDF.

3   U.S. Congress, House, Committee on Interior and Insular Affairs, Subcommittee on Energy and the Environment, Reporting of Information Concerning the Environment and Public Safety in Connection with Nuclear Facilities: Hearings Before the Subcommittee on Energy and the Environment, 95th Cong., 2nd sess., September 13, 1978 (Washington, DC: U.S. Government Printing Office, 1978), 131, https://www.google.com/books/edition/Reporting_of_Information_Concerning_the/FHyUmByQLuMC.

4   Steve Wing, David Richardson, Donna Armstrong, and Douglas Crawford-Brown, "A Reevaluation of Cancer Incidence Near the Three Mile Island Nuclear Plant: The Collision of Evidence and Assumptions," *Environmental Health Perspectives* 105, no. 1 (1997): 52–57, accessed June 28, 2025, https://www.ncbi.nlm.nih.gov/pmc/articles/PMC1469835/.

5   Aaron M. Datesman, "Radiobiological Shot Noise Explains Three Mile Island Biodosimetry Indicating Nearly 1000 mSv Exposures," *Scientific Reports* 10, 10933 (2020), accessed July 1, 2025, https://www.nature.com/articles/s41598-020-67826-5.

1.3: HealthCare.gov

1   Julie Appleby, "Former HHS Head Offers His Take on Health Law's Problems," *KFF Health News*, April 23, 2014, https://kffhealthnews.org/news/michael-levitt-former-hhs-qna/.

2   Henry Chao, *Success or Failure?: The Untold Story of Healthcare.Gov* (Advantage Media Group, 2018).

2.1: Crisis Engineering Toolkit: An Overview

1. Deborah Ancona, "Sensemaking: Framing and Acting in the Unknown," in *The Handbook for Teaching Leadership: Knowing, Doing, and Being*, ed. Scott Snook, Nitin Nohria, and Rakesh Khurana (SAGE Publications, 2012).

2.2: Establish a Crisis Engineering Center

1. See, for example, Anne Marthe van der Bles et al., "The Effects of Communicating Uncertainty on Public Trust in Facts and Numbers," *PNAS* 117, no. 14 (2020): 7672–7683, https://doi.org/10.1073/pnas.1913678117.

2.3 Map the System

1. Anthony Downs, *Inside Bureaucracy* (Little, Brown, 1967), chap. 12.
2. See, for instance, chapter 15 ("Search Problems in Bureaus") in Downs, *Inside Bureaucracy*.
3. "How TD Became America's Most Convenient Bank for Money Launderers," Bloomberg, March 18, 2025, https://www.bloomberg.com/news/features/2025-03-18/the-criminal-money-laundering-scams-that-cost-td-bank-billions.

2.4: Find Your People

1. Letter from Theodore Roosevelt to William Allen White, July 22, 1906, Theodore Roosevelt Papers, Library of Congress Manuscript Division, Theodore Roosevelt Center, Dickinson State University, https://www.theodorerooseveltcenter.org/Research/Digital-Library/Record?libID=o196144.

2.5: Take Novel Actions

1. John Allspaw, D. D. Woods, and Richard Cook built a consulting business around the thesis, called Adaptive Capacity Labs.
2. See, for instance, D. D. Woods, "STELLA: Report from the SNAFUcatchers Workshop on Coping with Complexity" (Columbus, OH: The Ohio State University, 2017).
3. Jennifer Mace, "Generic Mitigations," *O'Reilly*, December 15, 2020, https://www.oreilly.com/content/generic-mitigations/.
4. Robert B. Cialdini, *Influence* (Harper Business, 2006), 8.
5. Cialdini, *Influence*, 53.
6. Richard I. Cook, "How Complex Systems Fail," https://how.complexsystems.fail/#10.
7. Vilfredo Pareto, *Cours d'économie politique* (Lausanne, FR: F. Rouge, 1896).

2.6: Manage the Story

1. Paul Watzlawick, John H. Weakland, and Richard Fisch, *Change: Principles of Problem Formation and Problem Resolution* (W. W. Norton, 1974). As usual, we are paraphrasing to make the ideas fit in the current context, but we think the spirit is the same.

2. CNN compiled dozens of citations. See Daniel Wolfe and Daniel Dale, "'It's Going to Disappear': A Timeline of Trump's Claims That Covid-19 Will Vanish," CNN, October 31, 2020, https://edition.cnn.com/interactive/2020/10/politics/covid-disappearing-trump-comment-tracker/.

2.7: Measure Progress

1. Özlem Ergun et al., "Waffle House Restaurants Hurricane Response: A Case Study," *International Journal of Production Economics* 126, no. 1 (2010): 111–120, https://doi.org/10.1016/j.ijpe.2009.08.018.

2. Wilson, *Bureaucracy*, 161.

3. Downs, *Inside Bureaucracy*, 146.

4. Marina Nitze and Nick Sinai, *Hack Your Bureaucracy: Get Things Done No Matter What Your Role on Any Team* (Hachette Go, 2022).

5. Wilson, *Bureaucracy*, 162.

2.8: Communicate in a Crisis

1. Peter M. Sandman, *Responding to Community Outrage: Strategies for Effective Risk Communication* (American Industrial Hygiene Association, 1993); and The Peter M. Sandman Risk Communication Website, https://www.psandman.com/.

2. Peter M. Sandman, "Public Health Tells Noble Lies," The Peter M. Sandman Risk Communication Website, March 4, 2022, https://psandman.com/col/Corona64.htm.

3. Farrokh Alemi and Kyung Hee Lee, "Impact of Political Leaning on COVID-19 Vaccine Hesitancy: A Network-Based Multiple Mediation Analysis," *Cureus* 15, no. 8 (2023): e43232, https://doi.org/10.7759/cureus.43232.

4. "Dr. Peter M. Sandman: Crisis Communication (High Hazard, High Outrage)," The Peter M. Sandman Risk Communication Website, https://psandman.com/index-CC.htm.

5. Downs, *Inside Bureaucracy*, 112.

3.1: Know When You're Done

1. Downs, *Inside Bureaucracy*.

2. "Lowline Fire Update: Sept. 21, 2023," Incident Information System, accessed May 2025, https://inciweb.wildfire.gov/incident-publication/cogmf-lowline-fire/lowline-fire-update-sept-21-2023.

3.2: Spin Down the Crisis

1. Mary J. Waller and Seth A. Kaplan, *Crisis-Ready Teams* (Stanford Business Books, 2024).
2. Ruthanne Huising, "Moving off the Map: How Knowledge of Organizational Operations Empowers and Alienates," *Organizational Science* 30, no. 5 (2019), https://doi.org/10.1287/orsc.2018.1277.

3.3: Instill Change in the Wake of Crisis

1. Downs, *Inside Bureaucracy*, 139.
2. Nassim Nicholas Taleb, *Antifragile: Things That Gain from Disorder* (Random House, 2012), book 6.
3. Mace, "Generic Mitigations."
4. Frederick P. Brooks Jr., "The Second-System Effect," in *The Mythical Man-Month: Essays on Software Engineering* (Addison-Wesley Longman, 1975), 53–58.

3.4: Plan for a Future Crisis

1. We learned of this idea from Nassim Nicholas Taleb's *Incerto* series, particularly *The Black Swan*, 2nd ed. (Penguin Random House, 2010).
2. Deb Chachra, *How Infrastructure Works* (Riverhead Books, 2023), 161, Kindle.
3. Downs, *Inside Bureaucracy*, 130.
4. Downs, *Inside Bureaucracy*, 146.
5. Singer, *There Are No Accidents*, 5, Kindle.

3.5: Engineer a Crisis Career

1. Mark Walker, "FAA Audit of Boeing's 737 Max Production Found Dozens of Issues." *New York Times*, March 11, 2024, accessed June 28, 2025, https://www.nytimes.com/2024/03/11/us/politics/faa-audit-boeing-737-max.html.
2. Huising, "Moving off the Map."
3. United States Treasury, Office of Tax Analysis, "Number of People Who Have Ever Enrolled in ACA Marketplace Coverage, 2014–2024," September 3, 2024, https://home.treasury.gov/system/files/131/People-Enrolled-ACA-Mkt-Coverage-2014-24-09032024.pdf.

4.1: Ending California's Pandemic Unemployment Backlog

1. Lauren Hepler, "Internal Documents Reveal the Story Behind California's Unemployment Crash," *CalMatters*, November 7, 2023, https://calmatters.org/economy/2023/11/california-unemployment-covid/.
2. Lauren Hepler, "The Unemployment Money That Came Too Late," *CalMatters*, November 7, 2023, https://calmatters.org/economy/2023/11/unemployment-benefits-suicide-california-edd/.

Index

accountability sinks, 163–164
action bias, 145, 155
actuators, 122, 141–142
ad hoc adaptations, 246–247
ad hoc reporting, 202
adaptive capacity, 162–163, 203
advanced insights, 202
Affordable Care Act, 73. *See also*
    HealthCare.gov system
after-action reports, 238
alert fatigue, 208
anchoring bias, 33

Babcock & Wilcox, 56
backlogs, 251, 276–277, 289–292
bad news, hiding of, 61–64
basic monitoring, 201
batch processing, 166–167, 172
Beer, Stafford, 24
Benefit Systems Modernization
    project, 284, 297
biases, cognitive, 33–34, 61
Biden, Joe, 193
blame game, 38–39, 68
blocklists, 171
Boeing 747 cockpit design, 59, *60*
bracketing, 186
    cheat sheet, 194
    closing rituals, 192
    in Covid-19 pandemic, 192–193
    declaring endings at appropriate
        times, 189
    described, 185
    general techniques, 190–192
    investigating, 188
    noticing, 186–188
brownouts, 84
Bureaucracy (Wilson), 165
burndown chart, 292–295
burnout, 236–237
Burt, Gabriel, 75
business process, 128

CA EDD (California Employment
    Development Department) case
    study
    assessing if crisis exists, 276–277
    automated processes, 281–282
    backlogs, 277, 289–292
    Benefit Systems Modernization
        project, 284, 297
    burndown chart, 292–295
    call center, 280–281
    COVID-19 pandemic and, 275–276
    creating crisis engineering center,
        278–279
    crisis communications, 295–296
    dashboards, 298
    eliminating unnecessary
        duplicates, 287
    exited the crisis, 297
    expanding people's scope of action,
        286–287
    fraud prevention, 281–283,
        285, 297
    human-machine interface, 281–283

CA EDD case study (*cont.*)
  identification verification process, 281–283, 285, 297
  identifying experts/practitioners, 284
  manual processes, 282–283
  moving people around, 286
  points of resilience, 284
  progress measurement, 289–295
  self-service claim status tracker, 296
  status code conversions to plain English, 296
  stock and flow, 292
  storytelling, 289
  system mapping, 279–283, 298
  taking novel actions, 284–287
  transformative change, 297–299
  turning system off and on again, 285–286
  vendor dependencies and, 285
  work queue rules, 286–287
Centers for Medicare and Medicaid Services (CMS), 78, 86–87, 90
CGI Federal, 75
change agent, 266
Chao, Henry, 87, 91
CMS, 86
cognitive biases, 33–34, 61
cognitive dissonance, 120, 135, 204
Communicate in a Crisis, overview, 102. *See also* communications
communication channels, 109–110, 213
communications
  audience considerations, 212–215
  avoiding alert fatigue, 208
  avoiding surprises, 208
  avoiding unwanted/unhelpful actions, 208
  CA EDD case study, 295–296
  channels of, 109–110, 213
  characteristics of, 209–210, 218
  cheat sheet, 218–219
  costs of, 217niv
  cross-organizational, 251
  dashboards and, 212–213
  establishing guardrails, 217
  feedback, importance of, 216
  goals of, 208, 218
  improvement strategies, 215–216
  leveraging existing channels, 209–210
  low-latency, 109–110, 117, 279
  mobilizing the right people, 208
  to nontechnical colleagues, 214–215
  overview, 207
  in postcrisis mode, 251
  refining, 215–216, 218–219
  speakerphones, 109
  status page, 217
  targeted messaging, 213
  telling the truth, 210–212
  testing channels of, 259
  timeliness, 209
  of unpleasant information, 214
  updates, frequency of, 217
  winding down, 216–217, 219
complex systems
  batch-oriented vs. transactional, 137–138
  computers infiltrating, 19–20
  as cybernetic, 24
  defined, 21
  engaging with, 19
  examples of, 21
  as fractals, 22–23
  "how to turn it off" exercise, 136–139
  human to machine components ratio, 22
  machine-only systems, 22–23
  as misunderstood, 20–21
  purpose of, 24
  scatters operational responsibilities, 160
  testing, 136–139
complexity science, 21–22
components, 171–174
confirmation bias, 33
consensus reality, 13, 120–121
contractor, defined, 74nii

INDEX

contractor-government communications, 89
control loops, 122–124, 141–142, 262
controllers, 122, 132, 141–142
convening authority, 105–106, 116, 278
Conway, Melvin, 268
Conway's Law, 268
Cook, Richard, 177
core disruption, 15
COVID-19 pandemic, 192–193, 211–212, 275–276
crisis. *See also* future crisis; postcrisis regular order
  avoiding dangerous stories of, 38–41
  circumstances transformed into assets, 104
  common pitfalls, 38–41
  counterproductive action in, 288
  creating change, 301
  criteria, 276–277
  defined, 220–221
  destabilizing effects of, 17
  giving up, 102
  inverting power structures, 17
  manufacturing a, 220–222
  as necessary for rapid change, 64–65, 89–92
  outcomes after system enters, 38
  phony, 17
  readiness (*See* future crisis)
  reducing circumstances for, 229, 231
  scenario, 3–4
  as unavoidable, 301
  as useful, 36–37
crisis career
  acquiring missing skills, 271
  as a consultant, 270
  management hierarchy, 268
  for mid-career people, 271
  opportunities for, 267–268
  rest/recovery between projects, 270–271
  specialized role of, 269
  workload, 271

crisis cheat sheets
  communications, 218–219
  crisis engineering centers, 116–118
  crisis engineering team, 155–156
  crisis shutdown, 239–240
  future crisis, 264–265
  novel actions, 181–183
  postcrisis regular order, 253–254
  progress measurement, 205–206
  risk, 183
  storytelling, 194
  system mapping, 141–143
crisis communications. *See* communications
crisis engineer, 5, 23
crisis engineering
  building capacity for, 255–257
  in calm times, 256–257 (*See also* future crisis)
  defined, 4, 11–12
  defining success with, 26
  example of (*See* CA EDD (California Employment Development Department) case study)
  expense of active, 225–227
  goal of, 4
  overview, 301–302
crisis engineering centers. *See also* HealthCare.gov system
  announcement of, 112–113, 118, 279
  CA EDD case study, 278–279
  call to action, 114–115
  cheat sheet, 116–118
  closing ceremony, 235
  components of, 104
  convening authority, 105–106, 116, 278
  decision-making authority, 108–109, 117, 278
  designated venue for, 106–107, 116–117, 278
  events journal, 111–112, 117
  executive sponsor, 105
  future crisis readiness, 257–258

crisis engineering centers (*cont.*)
    HealthCare.gov system, 76
    identifying experts/practitioners, 110–111, 117
    incident leads, 110–112, 117
    kick-off ritual, 113–114, 118, 279
    low-latency communication, 109–110, 117, 279
    overview, 99, 103–104
    remote work and, 107
    shutting down (*See* crisis shutdown)
    unnecessary audience members, 115–116
crisis engineering team
    aftercare, 236–237
    burnout and, 236–237
    characteristics of, 145–147
    cheat sheet, 155–156
    communications and (*See* communications)
    composition of, 145
    developing new skill sets, 250
    disillusionment of, 237
    elevating gap fillers, 250
    expressing gratitude for, 234
    graduating out of team, 149
    identifying experts/practitioners, 110–111, 117, 147–148, 155, 284
    individuals with action bias, 145, 155
    introductions to, 148–149, 156
    invitations to join, 148
    new aspirations, 237
    nurturing new social connections of, 249–250
    points of contact outside of (*See* non-team members)
    postcrisis mode, 249–250, 254
    preexisting cliques, 149
    team size, 144–145
    transition to business as usual, 234
crisis indicators, 15, 276–277
crisis shutdown. *See also* postcrisis regular order
    after-action reports, 238, 240
    aftercare, 236–239, 240

cheat sheet, 230–231
closing ceremony, 235
ending check-ins, 234
engineering center and, 235–236
guidelines for, 227–230
lessons learned and, 238
mementos, 238
normalcy benchmarks, 228, 230
reintegrating teams, 234
returning to regular order, 226
reunion events, 238–239
rigidity cycle, 226
shortcut paths and, 233
signs of morbidity, 229–230
temporary approvals and, 233
timing of, 227
transition to business as usual, 232
turndown, 233–236, 239–240
turning off temporary infrastructure, 234–235
vendor relationships, 237
crisis toolkit, 100–101, 235
Cron jobs, 173–174
cross-functional project managers, 268
CSV files, 172
cultural failures, 59–61, 87–89
cybernetics, 23–25, 124

dashboards, 204, 212–213, 251, 298
data
    counteracting perverse incentives, 202
    human element to, 202, 206
    improving shared understanding of reality, 199–200
    in postcrisis mode, 250–251
    sensemaking with, 198–202, 205
Davies, Dan, 164
Davis-Besse Nuclear Power Station, 62–64, 262
deadlines. *See* rigid deadlines/timeframes
decision-making
    authority for, 108–109, 117

impediments to, 108
jury trials, 32–33
in postcrisis mode, 251
reduction in, 16
slowness in, 15
steps to, 32
System I systems, 34, 37–38
decision-making authority, 108–109, 117, 278–279
deferred/ignored maintenance, 252nii
degradation, disruption. or complete change of core processes or outcomes, 14–15, 276
degraded operations, 171
designated venue, 106–107, 116–117, 278
disillusionment, 237
distortion-resistant reporting, 260, 265
divergence behavior, 245
Dodge, Wag, 157
Downs, Anthony, 108niii, 126, 204, 226, 260
downtime, planned, 171

emergency core cooling system (ECCS), 48
events journal, 111–112, 117
exchange operations center. *See* HealthCare.gov system
executive sponsor, 105
executives. *See* management and executives
Extract-Transform-Load (ETL) tools, 171–172

failure of sensemaking, overview, 13–14, 276. *See also* sensemaking
failures, rationalized, 38
fast/no-gates release process, 172
features, turning off, 173
feedback, 165, 216
financial risk, 178
Find Your People, overview, 100, 110–111, 117, 284. *See also* crisis engineering team

flexibility, 16, 259, 289. *See also* resilience
Fowler, Martin, 172
fractals, 22–23
fraud prevention, 281–283, 285, 297
Fukushima Daiichi, 262
fundamental attribution error, 34
fundamental surprise, overview, 13, 276
future crisis
building crisis engineering capacity, 255–256, 264
cheat sheet, 264–265
crisis engineering centers and, 257–258
distortion-resistant reporting, 260, 265
gray swan events, 257, 264
multiple time scales, practicing, 261–262
preparation for, 257–258
prevention hygiene, 258–259, 265
readiness, 255–256
risky events, 257, 264
simulations, 259
system maintenance, 259
testing communications channels, 259
thoughtful preparation, 263
warning signs, 260

Garfinkel, Harold, 32
generic mitigations, 170
Gershman, Greg, 75
Goodhart, Charles, 204
Google, 86–87, 269
government culture, 88
gray swan events, 257, 264
Gresham's Law, 198
group sensemaking, 92–93, 262

Hastie, Reid, 33
health benchmarks, 196–197, 228, 248, 258

HealthCare.gov system
  abstractions leak, 82
  accelerated hiring, 91–92
  brownouts, 84
  change control board, 78–79
  CMS's lack of experience, 86–87
  complex contract structures, 76
  contractor-government communication, 89
  crisis engineering center, 76
  cultural failures, 87–89
  debugging efforts, 82
  evaluating situation, 74–76
  exchange operations center, 77–78
  experience, lack of, 86
  failed launch of, 73–74
  as a fractal, 22
  incident management, implementation of, 80
  mechanical failures, 81–82
  New Relic and, 92
  novel actions in, 161
  operator errors, 83–85
  organizational failures, 85–87
  overcomplicated configuration options, 81–82
  overlearned lessons, 85
  problem-solving capabilities, 92–93
  rescue of, 74–80
  responsibilities, diffuse, 85–86
  restructuring contracts, 90
  shared view of system performance, 76–77
  system integration, 81
  systems integrator contract, 90
  technical direction letters, 90
  as transactional website, 86–87
  troubleshooting, 84–85
  Vast Majority Sunday, 79–80, 93–94
  XOC incident lead, 79
high visibility, overview, 15, 276–277
high-pressure injection (HPI) pumps, 48
hindsight bias, 33
horizontal project managers, 268

Huising, Ruthanne, 269
human controllers. *See* controllers
human impact risk, 178
human-machine interface, 22, 123, 281–283

identification verification process, 281–283, 285, 297
incentives, 36
incident command. *See* crisis engineering centers
incident leads, 110–112, 117, 127–128
intuition, 186, 187
iterations, 180

jury trials, 32–33

Kahneman, Daniel, 33–34

Law of Counter-Control, 126
Law of Requisite Variety, 24–25
Layer Aleph, 7, 168, 191
legacy system excuses, 39–40
Leveson, Nancy, 55
load balancing/traffic routing, 171
logical persuasions, 36
low-latency communication, 109–110, 117, 279

Mace, Jennifer, 170
machine controllers. *See* controllers
machine-human interface, 22, 123, 281–283
Maclean, Norman, 14
maintenance resources, 252
Manage the Story. *See* storytelling
management and executives, 8, 114, 125, 132, 159, 260, 267–268
management hierarchy, 268
Mann Gulch wildfire, 14, 157
mapmaking, as social process, 134–135, 142–143. *See also* system mapping
McKenzie, Patrick, 282niv
Measure Progress, overview, 101

INDEX

measuring progress. *See* progress measurement
mechanical failures, 55, 81–82
Microsoft, 71

NASA Plum Brook Station, *45*
National Highway Traffic Safety Administration (NHTSA), 186
nested stories, keeping track of, 186
New Relic, 92
next-generation projects, 244–245
noncrisis environment, crisis engineering tactics in, 220–222
noncritical features, 172
nonregular order, 106
non-team members
  celebrating achievements, 152
  communications with, 154
  informal networks and, 152
  mitigating risk perceptions, 153–154
  overview, 156
  primary sources, 150–151
  protecting sources, 151
  reservoirs of resilience, 150
  responsiveness of, 151
  sharing information with, 154
normalcy benchmarks, 228
noticing
  anomalies, 188
  cheat sheet, 194
  defined, 186
  determining significance, 187–188
  looking for opportunities, 189–190
  overview, 184–185
novel actions
  accountability sinks, 163–164
  adaptive capacity, 162–163
  adding automation, 167
  adding/removing friction, 166
  ambiguity and, 162
  batch-processing and, 166–167
  best first, 158
  CA EDD case study, 284–287
  changing load on components, 171
  cheat sheet, 181–183

  earning buy-in for, 165
  eliminating/changing steps, 166
  enabling self-service, 167
  enumerating routine machine operations, 159–160
  example of, 157, 161, 168
  exploiting easy, 161–162
  feedback and, 165
  human capabilities, 162
  ideas for, 166–168
  identifying experts/practitioners, 174–175
  key stakeholder support, 165
  machine capabilities, 162
  Mann Gulch wildfire, 157
  modifying component input/output, 171–172
  modifying component tools, 172–173
  moving people around, 167
  one-line code change, 158–159
  open to new information, 176
  opening moves, 158–161, 181
  overview, 100–101, 157–158
  pilot/test, 165
  pivot, 176
  redundant/useless steps and, 169
  rehiring retired employees, 167–168
  rewriting in plain language, 167
  risk analysis (*See* risk)
  temptation to resolve everything, 175–176
  tools for, 169–174, 182
  trying processes in a new way, 166
  turning system off and on again, 158, 166
  unintended consequences, 180–181
  unleashing pending actions, 167
nuclear reactors. *See also* Three Mile Island accident
  control loops, 44, 122–124, 141–142, 262
  mechanics of, 43–44
  pressurizer, 44
  primary cooling loop, 43–44

nuclear reactors (*cont.*)
  secondary loop, 44
  on U.S. Navy ships, 56–57
  water hammer, 44
Nuclear Regulatory Commission, 47nvii

online transactional systems, 137–138
operator errors, 55–56, 83–85
organizational context, 129
organizational failures, 56–58, 85–87
overtraining, 39

Panchadsaram, Ryan, 75
Pareto Principle, 180
Park, Todd, 74
pending actions, 167
Pennington, Nancy, 33
Pennsylvania House of Representatives, 68–69
Perrow, Charles, 55
pilot-operated, defined, 47nvi
pilot-operated relief valve (PORV), 47
plain language, rewriting in, 167
plausibility, 298–299
plausibility beats accuracy, 93–94
pluggable pipelines, 171–172
political risk, 177
postcrisis regular order
  ad hoc adaptations, 246–247
  avoiding reimagining technical changes, 247–248
  behaviors to preserve, 245–248, 253
  cheat sheet, 253–254
  codifying key health metrics, 248
  creating stop/change/start opportunities, 242–243
  crisis engineering team and, 249–250, 254
  data collection, 250–251
  divergence behavior, 245
  documenting mitigations, 247
  elevating gap fillers, 250
  ending next-generation projects, 244–245
  formalizing new practices, 246
  instilling change in, 242–245
  maintaining artifacts created during crisis, 248
  returning to state of stability, 242
  rigidity cycle and, 242
  second-system effect and, 248
  shadow processes and, 246
  transformative changes in, 243–244
  unwanted crisis behaviors and, 249
progress measurement
  automating repeatable processes, 201–202
  CA EDD case study, 289–295
  cheat sheet, 205–206
  comparing expectations against outcomes, 200–201
  control via observation, 203–204
  creating measurements, 198–199
  dashboards and, 204
  defining goals, 196–197
  gauging severity of a disaster, 196–197
  Gresham's Law, 198
  improving shared understanding of reality, 199–200
  making a problem visible, 204
  overview, 195–196
  sensemaking with data, 198–202, 205
  setting health measures, 196–197, 205
  viewing progress, 204–205
pull notifications, 209

rationalized failure, 38
redundancy, 44, 169, 258
regular order, 106
remote work, 107
repetitive tasks, 173–174
reporting mechanisms, 260
repository, 159niii
reputational risk, 178

INDEX

resilience, 150, 178, 252, 258, 265, 271, 284
retired employees, 167–168
rigid deadlines/timeframes, 16, 277
rigidity attitude, 108niii
rigidity cycle, 226, 242, 253
risk
    balancing rapid action and strategic thinking, 176–177
    cheat sheet, 183
    financial, 178
    human impact, 178
    identifying, 178–179
    of inaction, 178
    mitigation, 153–154, 179–180
    overview, 176–177
    political, 177
    reputational, 178
    simulating, 178–179
    technical, 178
    types of, 177–178
rule-following acts, 108niii

Sandman, Peter, 211–212
Scranton, William, III, 61–62
scripting languages, 173
second-system effect, 248
self-delusion, 13
self-service, 167
sensemaking
    as active process, 298
    application of, 98
    cocreating your environment, 66
    as continuous, 67–69, 298
    defined, 4, 34–35, 98
    disadvantages, 31
    driven by plausibility, 70–71
    mechanics of, 36
    motivations for, 65, 297
    plausibility and, 298–299
    properties of, 35–36
    as retrospective, 65, 111, 297
    role in a crisis, 36–38
    as social process, 67, 92–93, 102, 144, 298

Sensemaking in Organizations (Weick), 35
sensors, 122, 141–142
shadow processes, 246
significance, overview, 186–188
Smith, Paul, 75
spreadsheets, 173
stakeholders, 165
standardized metrics, 202
static content, 172
status reporting/updating, 251–252
stock and flow, 292
story model, 33
storytelling
    accomplishing opposite of what you want, 191
    CA EDD case study, 289
    cheat sheet, 194
    closing rituals, 192
    curiosity creating investigation, 188
    declaring endings, 189 (*See also* bracketing)
    determining significance, 187–188
    disregarding factors you cannot change, 191–192
    getting unstuck, 190–192, 194
    looking for opportunity targets, 189–190
    noticing and, 186–188
    overview, 101
    relabel elements, 191
    reverse good and bad, 190–191
strangler pattern, 172
surprises, learning from, 37, 88
System I decision-making systems, 34, 37–38
system mapping
    CA EDD case study, 279–283, 298
    cheat sheet, 141–143
    completion of, 139
    control loops, 122–124
    cybernetics approach, 124
    disconnects, 134
    distortion-resistant questions, 132–133

system mapping (*cont.*)
  effective strategies, 129–130
  incident lead as owner of, 127–128
  as living, breathing tool, 131
  machine-human hybrids, 123
  mapping information flows, 127–131
  overview, 100
  plausibility, 121
  questions to ask, 128–129
  refining, 139
  self-encapsulating systems, 125–126
  as tool for transformative change, 134–135
  tracing end-to-end flow, 121–122
  uncovering messy, human realities, 130–131
  upstream controllers, 125–126
  value of, 140
  variations, identifying, 280
  widening scope of, 124–125
systems. *See* complex systems

tabletopping, 178–179
Take Novel Actions. *See* novel actions
technical leaders, 9
technical risk, 178
technical systems, 129
Three Mile Island accident
  as AI data center, 71
  blame game, 68–69
  cancer rates and, 70–71
  communication errors, 61
  control room design, 58, 59–60
  cultural failures, 59–64
  decay heat, 47–48
  diagram of, 46
  downplaying problems, 61–62
  emergency core cooling system, 48
  grid operators, 47, 49
  group sensemaking, 261–262
  high-pressure injection pumps, 48, 49
  human-machine interface, 59
  mechanical failures, 55
  Microsoft and, 71
  operator errors, 55–56
  organizational failures, 56–58
  pilot-operated relief valve, 47, 48, 57–58
  public understanding of the problem, 62–63
  reality vs. operator perception, 50–54
  reverberations of, 68–69
  scenario, 45–90
  scram, 47–48
  secondary loop feedwater pumps, 45–47
  sensemaking at, 64–72
  water lever measurements, 57–58
time scales, 261–262
timing constraints. *See* rigid deadlines/timeframes
TMI-2 nuclear reactor. *See* Three Mile Island accident
transactional systems, 137–138
transformative change, 134–135, 243–244, 253, 297–299
Trump, Donald, 192–193

Unaccountable Machine, The (Davies), 164
United States Digital Service (USDS), 80
U.S. Department of Labor, 283
U.S. Navy nuclear reactors, 56–57
user experience, 128

vendor dependencies, 285
vendor relationships, 237

Waffle House, 196–197
war room. *See* crisis engineering centers
Weaver, Matthew, 78, 80
Weick, Karl, 34–35
Wilson, James Q., 165
workforce scale up, 288

Zients, Jeff, 77, 79, 93

About the Authors

Marina Nitze's specialty is solving big, painful problems that others would rather avoid—particularly those involving a backlog. As the former chief technology officer of the U.S. Department of Veterans Affairs, Marina established the first agency-based USDS team, focused on improving veterans' access to care and benefits. Her team's effort helped increase veteran trust in the agency by more than 25 percent. Marina has been at the forefront of multiple hundred-million-dollar IT rescue and transformation projects. These efforts often went beyond restoration and propelled agencies into the future with processes like instant claims processing. Prior to the VA, Marina was a senior advisor on technology in the Obama White House, and the first entrepreneur-in-residence at the U.S. Department of Education. Marina is particularly passionate about improving America's foster care system. In her personal time, she runs TaskTackler, the personal productivity app for Type-A personalities. Marina previously authored the books *Hack Your Bureaucracy* and *Business Efficiency for Dummies*.

Matthew Weaver was pulled out of a first-grade public schoolroom, taught to program, and has never stopped trying to make larger and more complex systems work. Along the way he built rural wireless networks, and helped establish the site reliability engineering discipline at Google while responsible for web search operations there. Later, he was recognized by Fast Company for his role in building the first agency Digital Service team for the U.S. government after helping to rescue HealthCare.gov. In the government, he led more than twenty technical interventions across eleven federal agencies. His leadership was focused on some of the largest complex systems in existence, having budgets ranging from $100 million to $1.2 trillion, with half of them in the Department of Defense. After leaving the government, he founded Layer Aleph to use his unique skills and experience in service of his fellows. In his personal time he seeks strength and harmony as part of a community in rural America, along with his wife and dog.

Mikey Dickerson helped lead the HealthCare.gov rescue, for which he was featured on the cover of *Time* magazine. Afterward, President Barack Obama appointed him deputy chief information officer of the United States. Part of this involved Mikey being charged with establishing the United States Digital Service (USDS) to bring America's top technologists into government and solve its hardest and most pressing IT challenges. In addition to recruiting hundreds of developers and designers to join the federal government for tours of duty, he has led the successful transformation of a major, multi-million-dollar IT project at nearly every federal agency. Prior to USDS, Mikey served as one of the earliest site reliability engineers (SREs) at Google. There, he spent eight years growing and managing a team of SREs—the kind of engineer responsible for ensuring one of the world's most ubiquitous websites is safe and functioning, 24/7/365.